Petroleum Refining

FOR THE

NONTECHNICAL PERSON

SECOND EDITION

Petroleum

Refining

FOR THE
NONTECHNICAL PERSON

.

William L. Leffler

PennWell Books
PennWell Publishing Company
Tulsa, Oklahoma USA

Copyright © 1979, 1985 by
PennWell Publishing Company
1421 South Sheridan Road/P.O. Box 1260
Tulsa, Oklahoma 74101

Library of Congress Cataloging Publication Data

Leffler, William L.
  Petroleum refining for the non-technical person.

  Includes index.
  1. Petroleum—Refining. I. Title
TP690.L484   1984   665.5'3   84-26634
ISBN 0-87814-280-0

TABLE OF CONTENTS

FOREWORD

My duty is to speak; I have no wish to be an accomplice.
—I 'Accuse! I 'Accuse! Emile Zola

For as long as I have been associated with the oil industry, which is now about 20 years, I have been aware of the lack of tools available to the non-engineers to permit self-improvement. It was as if there were a technocratic conspiracy: "Let's keep the technology to the technicians!" Or maybe there is a caste system in oil companies. Those with the engineering degrees don't need a primer on technology; those without the technical degrees don't have a "need to know," so no technician attempted to articulate the *fundamentals*.

My experience is that both situations probably exist from time to time and in varying amounts, so I took pen and ruler in hand to combat the first and disprove the second. This book is intended to satisfy the needs of the young, astute, aspiring industry entrant who has no degree in chemical or process engineering or one of the closely related disciplines. It will be useful to the many persons in financial work, supplies, transportation, distribution, public relations, advertising, sales, or purchasing in petroleum-based or related companies.

The format is designed to permit little gulps at a time, followed by some reinforcing exercises which often build on previous work. If nothing else, by the time the reader is finished with the book he will not be intimidated by technical jargon. At least he will be able to ask more penetrating questions And that's most of the self-improvement battle won.

ADDENDUM TO THE SECOND EDITION

Technology doesn't sit still—nor could I. This second edition contains important additional material, particularly the chapter on simple and complex refineries. Some minor technical errors have been eliminated (happily very few were discovered in the first edition). Conversations with readers of the first edition have led to some rewrites of difficult material.

A word or two about help is in order. Linda Johnson typed the first edition of this book, and I never acknowledged it then. And that was before word processors! Beverly Mancuso has done the second edition, a painstaking job considering my weeny handwriting.

Bob Loughney, from his vantage point of third generation in the oil patch, has given me the anecdotes about the origins of the "sweet" and "sour" nomenclature and the 42-gallon barrel, confirmed my suspicions about API gravity and degrees Fahrenheit, and explained why hydrocarbons sell in the U.S. by the barrel but internationally by the *tonne*. I thank him.

W. L. Leffler, 1985

Petroleum Refining

FOR THE
NONTECHNICAL PERSON

INTRODUCTION

The beginning is the most important part of the work.
—The Republic, Plato

If you're reading this book, you don't need an introduction to the subject. You're already into it. It's not likely that anyone would read this book unless he or she had to.

The layout of the materials is designed to satisfy three different needs. It can be used as a reference book, because there's a good table of contents in the front and a good index in the back. The book can be used as a text in a course on refining processes. The combination of lectures, reading, and problem solving should be very reinforcing. But because most people will not have the luxury of listening to a lecturer, the layout is primarily designed for personal study. With that in mind, the rather dry material has been moistened with as much levity and practicality as could be precipitated.

For personal study, the following work plan and comments might be helpful. Chapter II on crude is most important. Chapter III on distilling has a lot of mechanical detail that's not fundamentally important. Don't let it dismay you. The material on flashing (Chapter IV), cat cracking, alkylation, reforming, residue reduction, and hydrocracking are all important. Understanding the mechanics of the refinery gas plants (Chapter VII) is less so.

The chapter on gasoline blending can be the most fun, in a toolish sort of way, because it deals with something familiar yet mysterious: car engines. Probably, though, the most important chapter is the one added in this edition, Chapter XIX, on simple and complex refineries. That chapter ties it all together because the difference between the simple and complex is the add-ons discussed in Chapters III through XVII.

Chapters XIII through XVIII, XX, and XXI are like *lagniappe,* a Cajun word meaning a small gift a merchant gives a customer at the time of sale. The information in those chapters is useful but not vital. So plan to keep your attention span through at least the first dozen chapters.

One final comment. This book should raise as many questions as you might have had before you started reading it . . . maybe more. That's not meant as an apology, but as a challenge! So find a process engineer and ask as you go along.

CRUDE OIL CHARACTERISTICS

Let these describe the undescribable.

—"Childe Harold's Pilgrimage," Lord Byron.

What is crude oil? The best way to describe it is to start by saying what it is not and what it doesn't do. It is not *a* chemical compound; it is a *mixture* of chemical compounds. The most important of its behavioral characteristics happens when it is heated up. When it is warmed to its boiling temperature and held there, it will not all evaporate.

Contrast that with water to make a point. Take a pot of water and heat it

Fig. 2-1 —Boiling Temperature of Water is 212° F.

3

to 212°F (Fahrenheit) and keep the heat on. What happens? The water starts to boil (to flash or vaporize). Eventually, if the heat is kept on, all the water will boil off.

If you had a thermometer in the pot, you would notice that the temperature of the water just before the last bit boiled off would still be 212°F. That's because the chemical compound H_2O boils at 212°F. At atmospheric pressure it boils at no more or no less.

Crude Oil Composition

Now back to crude. Unlike water, crude is not *a* chemical compound but thousands of different compounds. Some are as simple as CH_4 (methane); some are as complex as $C_{85}H_{60}$. CH_4 and $C_{85}H_{60}$ are the chemist's shorthand for certain chemical compounds. The nomenclature will be explained more fully in a later chapter (to avoid getting bogged down so early). They are all generally combinations of hydrogen and carbon atoms, called hydrocarbons. The important characteristic is that *each* of these compounds has its own boiling temperature, and therein lies the most useful and used physical phenomenon in the petroleum industry.

Distillation Curves

To explain, take the same pot and fill it with a medium weight crude oil. Put the flame to it and heat it up. As the temperature reaches 150°F, the crude oil will start to boil. Now keep enough flame under the pot to maintain the temperature at 150°F. After awhile, the crude stops boiling.

Step two, raise the flame and heat the crude to 450°F. Again the crude starts boiling and after awhile it stops.

You could repeat the steps on and on, and more and more crude would boil off. What is happening (you may have guessed) is that the compounds that boil at a temperature below 150°F vaporized in the first step, the compounds that boil at a temperature between 150°F and 450°F vaporized in the second step, and so on.

What you are developing is called a *distillation curve,* a plot of temperature on one scale and the percent evaporated on the other. Each type of crude oil has a unique distillation curve that helps characterize what kinds of chemical compounds are in that crude. Generally the more carbon atoms

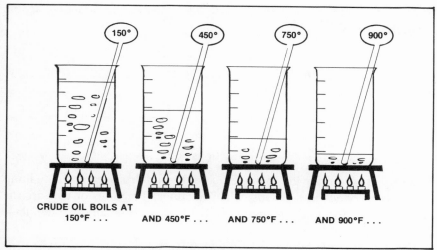

Fig. 2-2 —Boiling Temperatures of Crude Oil

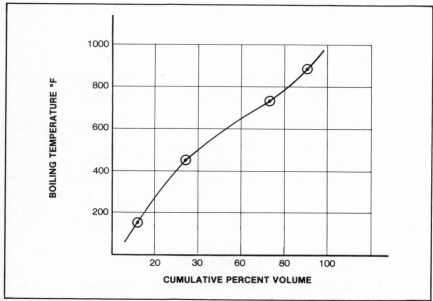

Fig. 2-3 —Crude Oil Distillation Curve

in the compound, the higher the boiling temperature, as shown in the examples below:

Compound	Formula	Boiling Temperature	Weight Pounds/Gallon
propane	C_3H_8	−44° F	4.2
butane	C_4H_{10}	31° F	4.9
decane	$C_{10}H_{22}$	345° F	6.1

Fractions

In further specifying the character of crude oil, it has been useful to lump certain compounds into groups called fractions. Fractions or cuts are the generic names for all the compounds that boil between any two temperatures, called *cut points*.

The typical crude oil has the following fractions:

Temperatures	Fraction
less than 90° F	butanes and lighter
90–220° F	gasoline
220–315° F	naphtha
315–450° F	kerosene
450–800° F	gas oil
800° F and higher	residue

Subsequent chapters will spend a lot of time discussing the characteristics of each of those fractions, but some are already apparent from the names.

It is important to note that crude oil compositions vary widely. The light crudes tend to have more gasoline, naphtha and kerosene; the heavy crudes tend to have more gas oil and residue. You may deduce from this that there is a relationship between the weight of the compounds and the temperature at which they boil. There is. Generally, the heavier the compound, the higher the boiling temperature. Conversely, the higher the cut points, the heavier the fraction.

Cutting Crudes

To pull together all this information on distillation curves, a little arithmetic manipulation may be instructive. Take the curves for the two

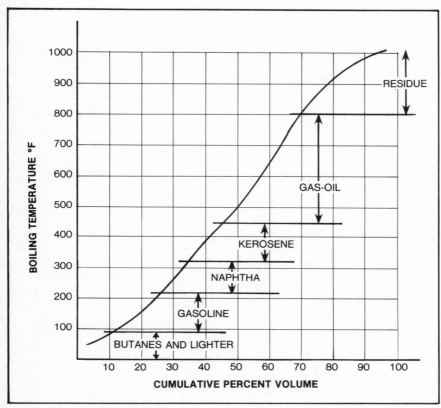

Fig. 2-4 —Crude Oil Distillation Curve and Its Fractions

crudes in Figure 2-5 and run through the steps to determine which crude has a higher kerosene content (a bigger kerosene cut).

Kerosene has a boiling temperature range from 315°F to 450°F.

Using Fig. 2-5 complete the following steps:

1. For the heavy crude, start from the vertical axis at 315° and intersect the distillation curve, going right, at point A. Point A is at 26% on the horizontal axis.

2. Now start at 450° and intersect the same distillation curve, going right, at point B. Point B is at 42% on the horizontal axis.

3. Calculate the cumulative percent volume from the initial boiling point

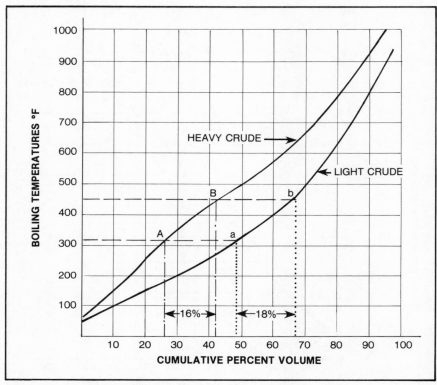

Fig. 2-5 —Kerosene Fraction in Two Types of Crude

of the kerosene to the end point: $42 - 26 = 16\%$. The heavy crude contains 16% kerosene.

4. Now do the same procedure for the light crude and find out that there is $66.5 - 48.5 = 18\%$.

Therefore the light crude has more kerosene in it than the heavy.

Gravities

Gravities measure the weight of a compound. Chemists always use a measure called *specific gravity* which relates everything to something universally familiar, water.

The specific gravity of any compound is equal to the weight of some volume of that compound divided by the weight of the same volume of water.

$$\text{Specific gravity} = \frac{\text{weight of the compound}}{\text{weight of water}}$$

The chemists' approach must have been too simple for the chemical engineer because the popular measure of gravity in the oil industry is a diabolical measure called *API gravity.* For some forgotten reason, the formula for API gravity, which is measured in degrees, is:

$$^\circ\text{API} = \frac{141.5}{\text{specific gravity}} - 131.5$$

If you play with the formulae a little bit, you'll find the following relationships, which might be the mental hooks on which you can hang the concepts.

1. Water has a specific gravity of 1 and an API gravity of 10°.
2. The higher the API gravity, the lighter the compound.
3. The reverse is true for specific gravity.

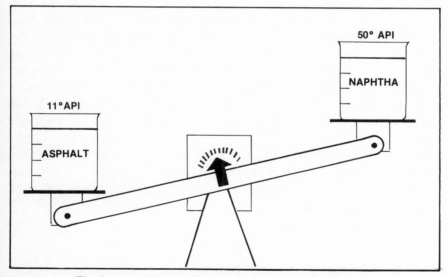

Fig. 2-6 —The Lower the API Gravity, the Heavier the Liquid

Typical gravities:

	Specific Gravity	API Gravity
Heavy crude	0.95	18°
Light crude	0.84	36°
Gasoline	0.74	60°
Asphalt	0.99	11°

Sulfur Content

A final divergence on the subject of crude oil quality is appropriate at this point—a discussion of sulfur content in crude. One of the annoying aspects Mother Nature endowed crude oils with is varying amounts of sulfur content in various types of crude oil. To complicate the endowment, the sulfur is not in the form of elemental sulfur but is usually a sulfur compound. That is, it is chemically bonded to some of the more complicated hydrocarbon molecules so that it is not easily separated from the pure hydrogen/carbon compounds.

The parlance in discussing crude oils of varying sulfur content is to categorize them into *sweet* crudes and *sour* crudes. This seemingly quaint, feintly oriental designation of sweet and sour has more to do with taste than you might think. In the early days of Pennsylvania crude oil production, petroleum was primarily a substitute for whale oil used as lamp oil for indoor lighting. If a kerosene fraction had too much sulfur, it would have an unacceptable smell when burned. The method originally used in the Pennsylvania oil fields to determine if the kerosene was suitable for shipping to the New York and Philadelphia markets was to taste it. If the taster thought it sweet, it passed; sour, it was rejected for having too much sulfur.

Today, sweet crudes typically have 0.5% sulfur or less, sour 2.5% sulfur or more. The area in between is sometimes called *intermediate sweet* or *intermediate sour*, but the distinction is not clear. What may be sweet to some may be sour to others, now that we have no more tasters.

Volumes

One other convention grew out of the Pennsylvania oil fields, according to oil-patch lore. Oil was initially shipped to market by wagon or flatcar in used 50-gallon wine barrels. To allow for spillage during transportation, payment at the destination was for only 42 gallons. Receivers still pay on that basis. Shippers soon learned to ship that way, too.

Actually, two standards developed for measurement. In the U.S., since transportation was primarily over land by wagon, train, or eventually pipeline, the measurement was easier by volume. In the rest of the world, particularly in Europe, most petroleum was transported by seagoing ves-

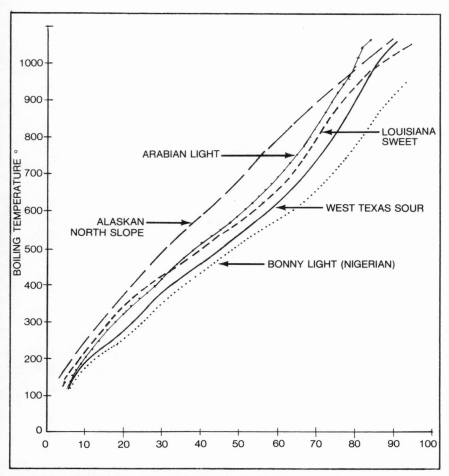

Fig. 2-7 —Distillation Curves for Some Crude Oils

sel. There, the easier method of measurement was weight (displacement). Hence, in the U.S. almost all commerce is done in barrels. In Europe, much is done in tonnes.

Review   In Fig. 2-7, the distillation curves for five different crudes are shown, three domestic and two foreign. Some have more light fractions, some have more heavy. All have different prices, so refiners will have different incentives to process them—particularly when the different hardware in the refineries is taken into account.

EXERCISES

1. (a) Draw the distillation curves for the following crude oils (on the same graph).

	% volume	
	Oklahoma sweet	*Heavy California*
lighter than 113°	5.1	–
113 to 220	9.2	–
220 to 260	4.0	–
260 to 315	5.7	4.2
315 to 390	9.3	5.1
390 to 450	5.4	4.8
450 to 500	5.8	8.5
500 to 550	4.7	7.9
550 to 650	10.8	8.0
650 to 750	8.6	14.8
750 to 900	13.5	15.1
900 to 1000	5.9	13.4
over 1000	12.0	18.1

(b) How much naphtha (220°-315°) is there in each crude?

2. Suppose you had a beaker of asphalt (11° API) and a beaker of naphtha (50° API), both equal volumes. If you mixed them together, what would the resulting API gravity be? The answer is not "30.5° API."

DISTILLING

Why should we rise, because 'tis light?

—"Break of Day," John Donne

A popular mistake by the casual passerby of a refinery is to refer to the many tall columns inside as "cracking towers." In fact, most of them are distilling columns of one sort or another. Cracking towers, which are usually shorter and squatter, will be covered in a later chapter.

Distilling units are the clever invention of process engineers who exploit the important characteristic discussed in the last chapter, the distillation curve. The mechanism they use is not too complicated but, for that matter, not that interesting. However, in the interest of completeness and familiarity, the rudiments will be covered here.

The Simple Still

An analogy will be useful. The Kentucky moonshiner uses the simple still to separate the white lightning from the dregs. After the sour mash has fermented, i.e., a portion of it has slowly undergone a chemical change to alcohol, it is heated to the boiling range of the alcohol. The white lightning vaporizes; as a vapor, it is less dense (lighter) than liquid. So it moves up, out of the liquid, and then through the condenser where it is cooled and turned back to liquid. What's left in the still is discarded. What goes overhead is bottled. This is simple batch process distillation.

If a moonshiner wanted to sell a better-than-average product, he might run his product through a second *batch still* much like the first, in order to separate the best part of the liquor from some of the nonalcoholic impurities that inevitably flow along with the overhead in the first still. They might have gone overhead because of his inability to control the temperature of the sour mash boil or because he wants to be sure he gets all he can and purposely sets the temperature a little high on the first batch.

Such a two-step operation could be made into the continuous operation

13

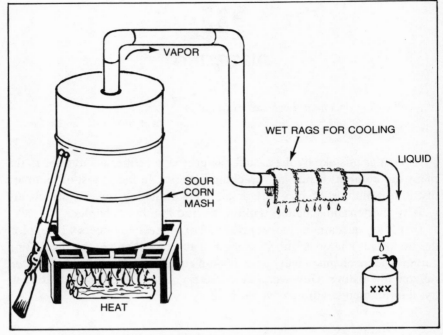

Fig. 3-1 —The Moonshiner's Still

shown in Fig. 3-2. In fact, many early commercial distilling operations looked like that.

The Distilling Column

The batch distilling operation above is obviously not suited for handling 100-200,000 barrels/day of crude oil with five or six different components being separated. The distilling column can do it on a sustained basis with much less labor, facilities, and energy consumption in the form of fuel/heat.

From afar, what happens at a crude distilling column is shown in Fig. 3-3. Crude goes in, and the products going out are the gases (butane and lighter), gasoline, naphtha, kerosene, light gas oil, heavy gas oil, and residue.

What goes on inside the distilling column is more subtle. The first

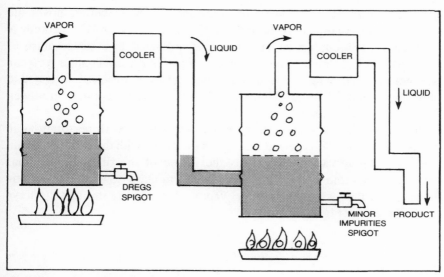

Fig. 3-2 —Two Stage Batch Still

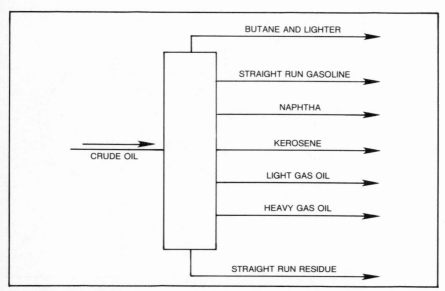

Fig. 3-3 —Distilling

element important to the operation is the *charge pump*, which moves the crude from the storage tank through the system (see Fig. 3-4). The crude is first pumped through a furnace where it is heated to a temperature of perhaps 750°F. From the knowledge developed in the last chapter, you can see that over half of the crude oil changes to the vapor form as it is heated to this temperature. This combination of liquid and vapor is then introduced to the distilling column.

Inside the distilling column there is a set of *trays* with perforations in them. The perforations permit the vapors to rise through the column. When the crude liquid/vapor charge hits the inside of the distilling column, gravity causes the denser (heavier) liquid to drop toward the column bottom, but the less dense (lighter) vapors start moving through the trays toward the top (Fig. 3-5).

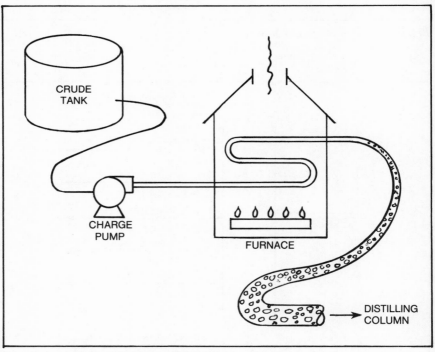

Fig. 3-4 —Crude Oil Feed to Distilling

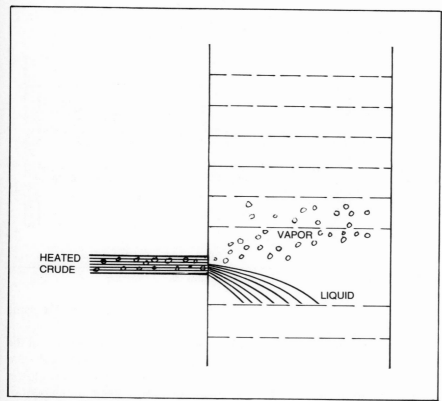

Fig. 3-5 —Crude Entering the Distilling Column

The perforations in the trays are fitted with a device called *bubble caps* (Fig. 3-6). Their purpose is to force the vapor coming up through the trays to bubble through the liquid standing several inches deep on that tray. This bubbling is the essence of the distilling operation: the hot vapor (starting out at 750°F) bubbles through the liquid. Heat transfers from the vapor to the liquid during the bubbling. So as the vapor bubbles cool a little, some of the hydrocarbons in them will change from the vapor to the liquid state. As heat transfers from the vapor to the liquid, the temperature of the vapor drops—the lower temperature of the liquid causes some of the compounds in the vapor to condense (liquefy).

After passing through the liquid and shedding some of the heavier

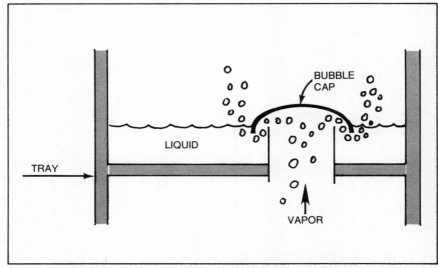

Fig. 3-6 —Bubble Cap on a distilling column tray

hydrocarbons, the vapor then moves up to the next tray where the same process takes place.

Meanwhile, the amount of liquid on each tray is growing as some of the hydrocarbons from the vapor are stripped out. So a device called a *downcomer* is installed to permit excess liquid to overflow to the next lower tray. Every so many trays, the liquid is drawn off to keep the amount of product going out of the distilling column equal to the crude coming in.

What happens is that some molecules make several round trips: up a couple trays as vapor, finally condensing, then down a few trays via the downcomer as a liquid. It's this vapor-liquid mutual scrubbing that separates the cuts. Once through won't do it.

At several levels on the column, the *sidedraws* (Fig. 3-7) take the liquid distilled product off—the lighter products from the top of the column, the heavier liquids toward the bottom.

Reflux and Reboil

There are several things that go on outside the distilling column that facilitate the operation. To assure that some of the heavies don't get out the

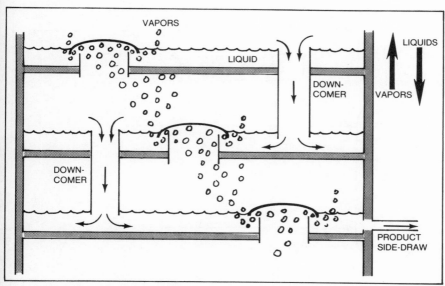

Fig. 3-7 —Downcomers and Sidedraws

top of the column, sometimes some of the vapor will be run through a cooler. Whatever is condensed is reintroduced to a lower tray. Whatever is still vapor is sent off as product. The process is a form of *refluxing* (Fig. 3-8).

Conversely, some of the light hydrocarbon could be entrained on the bottom of the column where the liquid part of the crude oil ended up. So a sidedraw may be used to recirculate the liquid through a heater to drive off any lighter hydrocarbons for reintroduction into the distilling column as a vapor. This is called *reboiling*. The advantage of a reboiler is that only a small fraction of the crude stream has to be worked on to accomplish the additional recovery. The whole crude volume doesn't have to be heated, saving energy and money.

Reboiling or refluxing can be used effectively in the middle of the column as well, facilitating good separation. The advantage of the reboiler is that it gives a heat input to help push lighter molecules up the column. Similarly, refluxing can give a last shot at condensing some heavy molecules that may have gotten too high in the column.

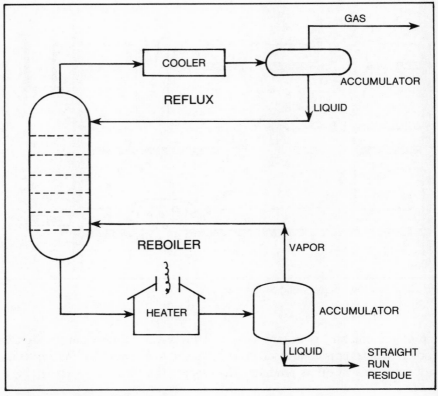

Fig. 3-8 —Reboil and Reflux

Sometimes the crude composition is such that there would not be enough liquid/vapor flow on some of the trays of the column. In that case, using reflux and reboil can regulate the flows to keep the distilling (separating) operation going.

Cut Points

To someone analyzing distilling operations, the key parameters are the *cut points*. The temperatures at which the various distilling products are separated are called the cut points. Specifically, the temperature at which a product (or cut or fraction) begins to boil is called the *initial boiling point*

(IBP). The temperature at which it is 100% vaporized is called the *end point* (EP). So every cut has two cut points, the IBP and the EP.

By referring to the diagram in Fig. 3-3 again, it becomes readily apparent that the end point of naphtha is the IBP of kerosene. At a cut point, the EP and the IBP of the two cuts are the same, at least nominally.

Depending on how good a separation job the distilling operation does, the IBP and the EP might not be the same. You may have wondered, looking at the mechanics of trays and bubble caps, how well the process works. In fact, the operation is a little sloppy, resulting in what is referred to as, pardon the expression, *tail ends*.

For instance, if the naphtha and kerosene were analyzed in a lab, the two distillation curves would look like the curves below. Looking closely, the end point of naphtha is about 325; the IBP of the kerosene is about 305.

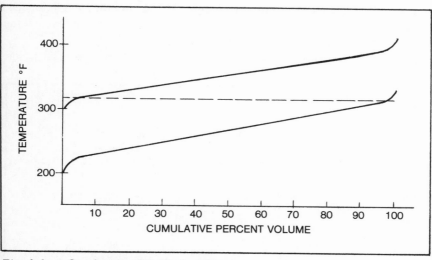

Fig. 3-9 —Overlapping Distillation Curves

A better way of visualizing the tail ends is in Fig. 3-10, which plots temperature—this time not against cumulative percent vaporized but against percent being vaporized at that temperature (the first derivative of the above diagram, if you know calculus).

Almost all refining operations have the phenomenon of tail ends. It is so common, it is taken for granted. However, to simplify analysis, an accommodation is made. The usual cut temperatures in any distilling operations are generally the *effective cut points*. These represent the compromise temperature at which the cuts can be considered cleanly cut. The balance of this book will implicitly mean *effective cut point* when the term cut point is used.

Setting Cut Points

The cut points in the last chapter and in the discussions above may have been taken as a given for each fraction, but there is some latitude in setting the cut points on a distilling column. Changing the cut point between naphtha and kerosene has several implications. If the cut point were changed from 315 to 325, several things would happen. First, the *volumes* coming out of the distilling column change—more naphtha, less kerosene. That's because the fraction between 315 and 325 now comes out the naphtha spigot instead of the kerosene spigot.

At the same time the *gravities* of both naphtha and kerosene get heavier.

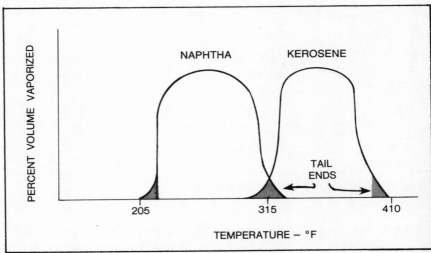

Fig. 3-10 —Tail Ends in Distillation Cuts

How can that be? The cut moving into naphtha is heavier than the average naphtha gravity. It is also lighter than the average kerosene gravity. So both cuts get heavier!

Some of the other properties also change, but gravity is the only one that has been discussed so far. When the downstream operations are discussed in later chapters, you'll appreciate the further implications of changing cut points on the distilling unit.

A listing of the destinations of all the distilling unit cuts will help set up future chapters. The light ends from the top of the column (the overhead) go to the gas plant for fractionation. The straight run gasoline goes to motor gasoline blending. Naphtha goes to the reformer for processing. Kerosene goes to a hydrotreater for cleanup. Light gas oil goes to distillate fuel blending. Heavy gas oil goes to the cat cracker as feed. And straight run residue is fed to the flasher.

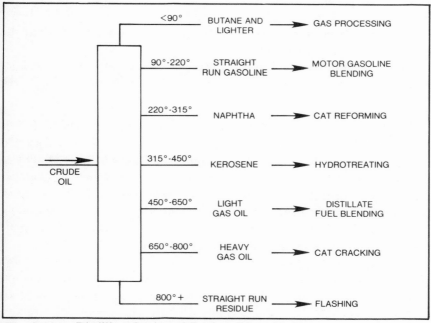

Fig. 3-11 —Distilling Crude and Product Disposition

EXERCISES

1. Fill in the blanks below with nomenclature from the following list:

tail ends	downcomer	overhead
IBP	EP	cooler
furnace	straight run gasoline	continuous
crude oil	fractionator	
batch	bubble cap	
increases	decreases	

   a. After moonshine comes out the top of a still, it *has* to be run through a _____ before it can be bottled.

   b. _____ processing is not very practical in a modern refinery. Today all crude distillation is a _____ operation.

   c. The device that facilitates the essential job of scrubbing in a distilling unit is the _____ _____.

   d. The holes in the distilling column trays are fitted with either a _____ _____ or a _____.

   e. Tail ends occur because the _____ of one cut overlaps the _____ of another.

   f. As vapor moves up the distilling column, its temperature _____.

   g. Lowering the end point of a distilling column cut _____ the volume of that cut and _____ the API gravity.

2. The refinery Operations Manager is told he must produce 33 MB/D of furnace oil this winter. He is also told he will be supplied with 200 MB/D of crude—30 M B/D of Louisiana crude and 170 M B/D of West Texas. The distillation curves for those crudes are shown below. Another premise in the operation is that he must maximize jet fuel. That is, he must squeeze as much out of the crude as possible. He knows that the boiling range for jet fuel is 300-525° F, so he will have to set the cut points on his distilling units at these two temperatures.

Finally, in order to meet the specifications for the 33 M B/D of furnace oil, 20 M B/D of straight run light gas oil (SRLGO) will have to be cut from the crude on the distilling units and provided to the furnace oil blending operation.

Problem: At what temperatures will the Operations Manager have to cut SLRGO to make 20 M B/D?

Distillation Data:

	West Texas	Louisiana
	% Volume	*% Volume*
IBP-133° F	15	20
113-260	12	18
260-315	18	15
315-500	10	15
500-750	20	12
750-1000	10	10
over 1000	15	10

Hints: Calculate a composite crude oil distillation curve. The turbine fuel EP is the SRLGO IBP. So calculate the SCLGO EP that gives 20 M B/D.

*Save the answer to this problem. Problems in subsequent chapters will build on this exercise.

FLASHING

A vacuum is repugnant to reason.

—Principles of Philosophy, Rene Descartes

When distillation curves and distilling columns were discussed in the last two chapters, the shape of the curve at temperatures around and above 900°F was purposely left vague. There is a phenomenon that happens at these temperatures called *cracking*. To accommodate this phenomenon and remedy its bad symptoms, the process of *vacuum flashing* was developed.

The Cracking Phenomenon

Suppose a laboratory technician were going to develop a crude distillation curve. He would heat the crude oil; note and record the temperature; capture the vapor and condense (liquefy) it; and record and cumulate the volume of the liquid. If he permitted the temperature of the crude to rise to about 900°F, a sudden change in the shape of the distillation curve would take place. As the temperature went from 900 to 1100°F, the cumulative volume recovered would exceed 100%, and crude would still be boiling in the pot.

This may be an exaggeration of what can be observed easily with the naked eye, but it makes the point. As the complicated hydrocarbon molecules that have not yet vaporized at 900°F are heated to higher temperatures, the energy transfer from the heat is enough to *crack* the molecules into two or more smaller molecules. For example, a molecule of $C_{16}H_{34}$ *may crack into three pieces*—C_8H_{18}, C_6H_{12}, and C_2H_4 as shown in Fig. 4-1. (A fuller discussion of the chemical reaction is cracking will be covered in the chapter on cat cracking.)

Recalling the discussion on boiling points, it becomes apparent why the shape of the distillation curve changes so abruptly. The smaller molecules boil at much lower temperatures than the larger ones. As they are created by cracking, they leap out of the boiling crude. But that is only half the

26

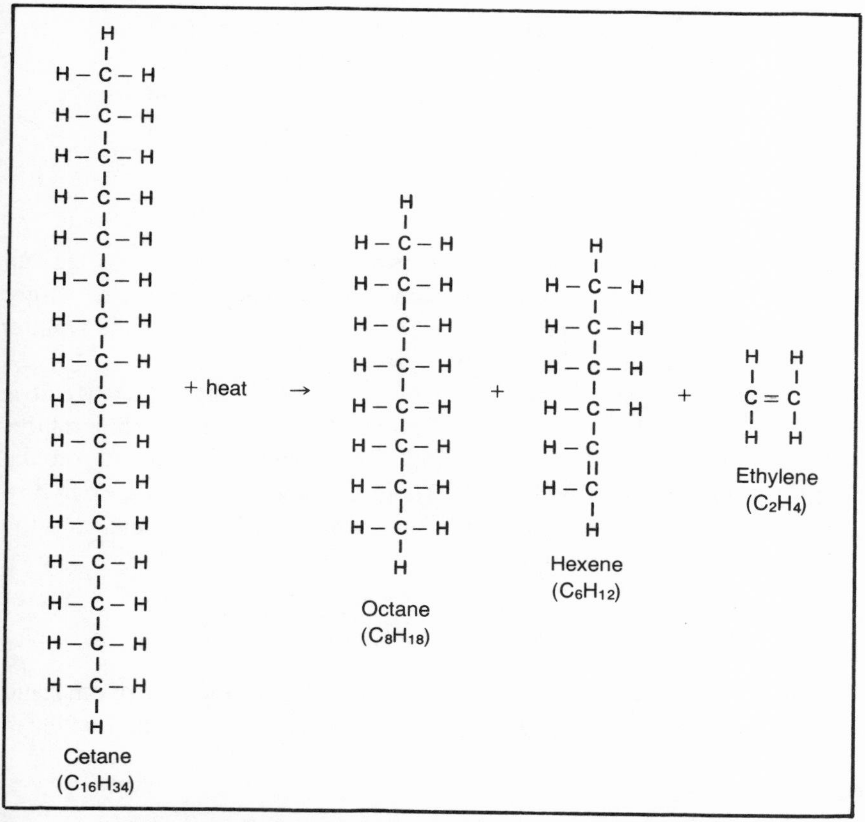

Fig. 4-1 —Molecule Cracking

story. Why does more volume get created? In simplest terms, the small molecules take up more room *per pound* than the large ones.

Again recall the discussion on hydrocarbon properties and specific gravities. The compound $C_{16}H_{34}$ weighs 7.2 lb/gal, but C_8H_{18}, C_6H_{12}, and C_2H_4 weight 5.9, 5.6, and 3.1 lb/gal, respectively. Take a gallon of $C_{16}H_{34}$ and suppose you could crack it completely into these three components. (It never happens that way; combinations of different compounds turn up, too.) The chemistry of it says that there would be 50% C_8H_{18}, 38% C_6H_{12}, and 12% C_2H_4. But the percentages are *by weight, not by volume*. If you start with 7.2 lb, you end up with 7.2 lb. How many gallons? Calculate them as follows:

$$7.2 \text{ lbs of } C_{16}H_{34} = 7.2 \text{ lb} \div 7.2 \text{ lb/gal} = 1.00 \text{ gal}$$
$$50\% \text{ yield of } C_8H_{18} = 3.6 \text{ lb} \div 5.9 \text{ lb/gal} = 0.61$$
$$38\% \text{ yield of } C_6H_{12} = 2.7 \text{ lb} \div 5.6 \text{ lb/gal} = 0.48$$
$$12\% \text{ yield of } C_2H_4 = 0.9 \text{ lb} \div 3.1 \text{ lb/gal} = 0.29$$

100%	7.2 lb	1.38 gal

So, 7.2 lb of $C_{16}H_{34}$ take up 1 gal, but 7.2 lb *of the components* take up 1.38 gal. The simple reason this phenomenon exists is that the larger molecules tend to have the atoms packed closer together than the smaller molecules. So they take up less space per pound.

Cracking is an interesting and lucrative process, but only when it is controlled. The distilling unit is not designed to control it, so temperatures that cause it are avoided in distilling. The heaviest cut points on the distilling column are around 750°F. There are, nonetheless, lots of hydrocarbons that need to be separated in the straight run residue cut, and the technique used is vacuum flashing.

Effects of Low Pressure

Suppose you had the hot dog concessions in both the Astrodome and Mile High stadiums. Do you know that, on the average, to boil water to

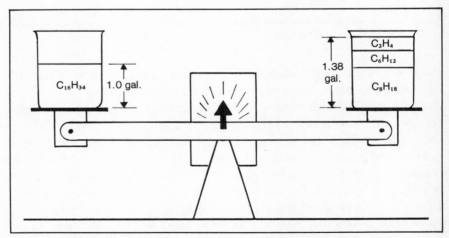

Fig. 4-2 —C_{16} H_{34} and its Cracked Components Weigh the Same

cook hot dogs you'd heat the water to 212°F in Houston but only 210°F in Denver? This temperature gradient represents the effect of different atmospheric pressures caused by change in altitude. When people refer to the air in the mountains being thinner, they really mean less dense, which is because the pressure is less.

The relationship between pressure and boiling temperature is simple. The process of heating only helps the molecules to absorb enough energy to escape from the liquid form. The rate at which they escape depends on the rate at which heat is delivered to them *and* the pressure of the air above them. The lower the pressure, the less energy has to be transmitted, and the lower the temperature at which vapor will start forming in the liquid, i.e., boiling.

The point is—the lower the pressure, the lower the boiling temperature.

Vacuum Flashing

Apply the pressure-boiling temperature relationship to the crude oil cracking problem. The straight run residue will crack if the temperature goes too high; the straight run residue still needs to be separated into more cuts. The solution is to do the fractionation at reduced pressure.

Straight run residue is pumped directly from the distilling column to the flasher. Because of the way distilling works, the temperature of the residue is equal to the IBP, with a couple of degrees to allow for cooling. The straight run residue is taken into a squat, large diameter column where the pressure has been lowered. Atmospheric pressure is 14.7 lb per square inch (psi), and that is about what distilling columns have inside. The pressure in a vacuum flasher is about 4.5 to 5.5 psi. At that reduced pressure, the lighter fractions of the residue will begin to boil (flash).

Flashing is a cooling process. That is why you stick your forefinger in your mouth, then hold it up in the air to find the wind direction. The breeze evaporates the liquid, causing one side of your finger to be cooler than the other. In order to counteract the cooling process, superheated steam (steam under pressure with temperatures above 750°F) is also introduced to the vessel. Heat is transferred from the steam to the residue, maintaining the temperature and the flashing. The other function of the steam is vacuum control. Pressure in the vessel is kept very low by a vacuum pump at the top of the vessel.

As shown in the diagram, several streams can be drawn off the flasher.

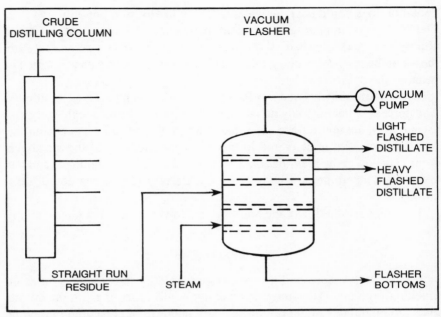

Fig. 4-3 —Vacuum Flashing

Light flashed distillate and *heavy flashed distillate* are sometimes kept segregated. Either can sometimes be used as lubricating oil feedstocks. Often they are not segregated but drawn off together and called *flasher tops*.

The heavy material from the bottom of the flasher is called, appropriately enough, *flasher bottoms* and is used as feed to an asphalt plant or a thermal cracker, or as a blending component for residual fuel.

Flashing is the equivalent of distilling the straight run residue at a cut point in the neighborhood of 1000-1100°F. Most distillation curves are depicted as if the theoretical distillation took place.

Because even vacuum flashing has its practical limitations, the end point of the straight run residue i.e., the end point of the crude oil, is never ascertained. Vacuum distillation is not run that high. This lack of knowledge is of no consequence, however, because the products using the bottom of the barrel are not sensitive to end point, per se. Other measures, such as gravity and viscosity, are the properties used to control the quality of the flasher bottoms.

EXERCISES

1. Fill in the blanks:
   a. Vacuum flashing is a technique to continue _____ of crude oil at high temperatures without encountering _____.
   b. The lower the pressure around a liquid, the _____ (lower, higher) the boiling temperature.
   c. The higher the pressure in a vacuum flasher, the _____ (lower, higher) the end point of the flasher tops, assuming the feed rate and feed temperature don't change.
   d. Products from a flasher are called _____, _____ and _____.

   e. To reduce the volume of flasher tops, holding the feed rate constant, you either _____ the pressure in the flasher or _____ temperature of the flasher feed.

2. Use the data given to the Operations Manager in the previous chapter. In order to satisfy the need for feed to the asphalt plant, 35 M B/D of flasher bottoms must be made available. (a) What is the maximum cut point at which the flasher must operate to assure supply to the asphalt plant? Use the distillation curve temperatures, not the internal flasher pressures and temperatures. (b) If the straight run residue was cut from the distilling column at 800°F, what would be the volume of flasher tops?

THE CHEMISTRY OF PETROLEUM

It strikes me that Nature's system must be a real beauty, because in chemistry we find that the associations are always beautiful whole numbers—there are no fractions.

—R. Buckminster Fuller

So far, in most of the discussion of crude oil and refining you have been able to avoid the basics of chemistry; but the good times have come to an end. The rest of the refining processes, for the most part, are chemical reactions. While you don't have to have taken Chemistry I and have gotten at least a C in Physical Chemistry, you will have to know something about hydrocarbons to go on from here.

Don't skip this chapter! You will have to keep a figurative thumb in these pages if you want a full appreciation of the subsequent seven chapters.

Atoms and Molecules

Physicists are continually surprising the scientific world by breaking matter into smaller and smaller particles, even more esoteric than neutrons and electrons. Luckily, the smallest notion in petroleum refining is the atom. Examples are carbon, hydrogen, sulfur or oxygen, whose chemical symbols are C, H, S and O.

The characteristics of matter depend on the types of atoms that make it up and how they are attached to each other in groups called molecules. There are rules by which atoms can be arranged into molecules. The most important rules have to do with valences and bonds.

Valences. Each type of atom (element) has an affinity for other elements according to its atomic structure. For example, carbon atoms would always like to attach themselves to four other atoms. Hydrogen atoms would always like to attach themselves to only one other atom.

Definition: The valence of an atom of any element is equal to the number of hydrogen atoms (or their equivalent) that the atom can combine with.

Bonds.   The connection between two atoms is called a bond. You can think of it as an electrical force that ties the two atoms together.

Hydrocarbons

The simplest example of valence, bonds and hydrocarbons is methane, which has one carbon and four hydrogens, CH_4. Looking at the structure, you can see all the valence rules are satisfied.

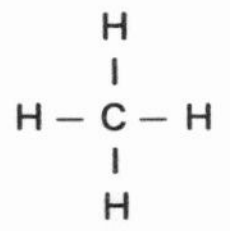

```
          H
          |
     H  – C  – H
          |
          H
       Methane
        (CH₄)
```

Fig. 5-1 —Methane

If the next simplest hydrocarbon, ethane, is chosen, some further complications can be introduced. Ethane is C_2H_6. You can check that each carbon has four attachees; each hydrogen has one. Note that one carbon atom is attached to another one. That is okay. As a matter of fact, whenever you see hydrogen referred to as a separate compound, it will always be called H_2 because that is how it exists—one hydrogen attached to a second, satisfying the valence rule of both atoms.

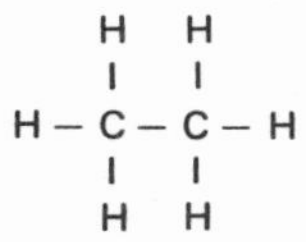

```
        H    H
        |    |
    H – C  – C – H
        |    |
        H    H
        Ethane
        (C₂H₆)
```

Fig. 5-2 —Ethane

A whole class of hydrocarbons can be defined by extending the relationship between methane and ethane. These molecules are *paraffins* or straight chain compounds and have the formula C_nH_{2n+2}. Examples are propane, normal butane and normal pentane.

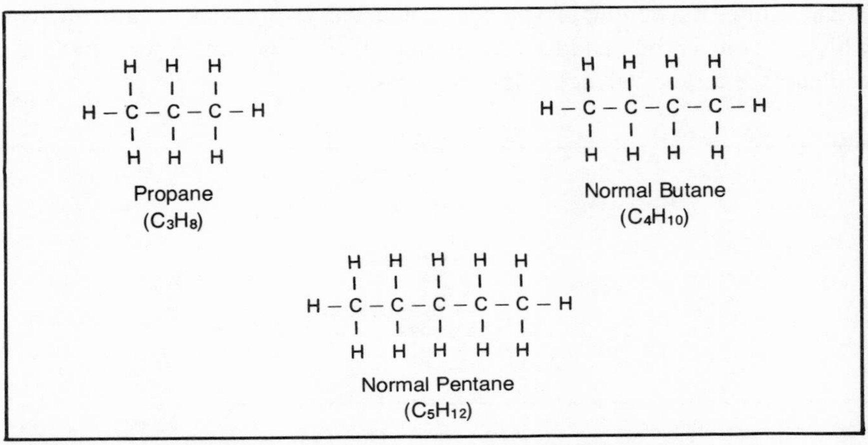

Fig. 5-3 —Propane, Normal Butane, Normal Pentane

Why the prefix "normal" before butane and pentane? Well, there are several ways to arrange the carbon atoms in C_4H_{10} and C_5H_{12}. One of them is shown in the figure above. Another is to put a *branch* off one of the inside carbons. In that case, the compounds C_4H_{10} and C_5H_{12} are called isobutane and isopentane.

Even though normal butane and isobutane have the same formula, they behave differently. They boil at different temperatures; they have different

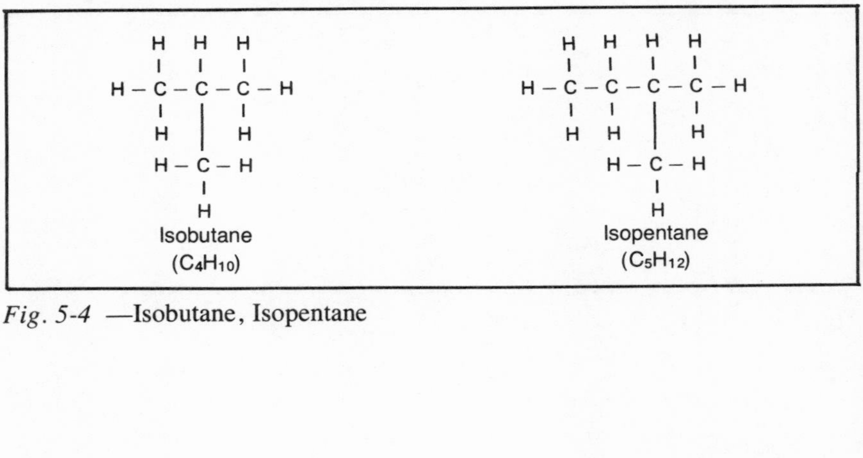

Fig. 5-4 —Isobutane, Isopentane

gravities (since they are packed differently); and they cause different chemical reactions which will be important in a later chapter on alkylation.

There is a shorthand convention used when talking about these light hydrocarbons. When talking about mixtures of streams that have, say, only propane, ethane, and maybe hydrogen but no butanes nor molecules larger (heavier), that mixture or stream could be called propane and lighter, or C_3-. Also, it could be said it contains no butane and heavier or C_4+. The convention is used for any of the light hydrocarbons through C_5.

Naphthenes

Another class of hydrocarbons, *ring* or *cyclo compounds,* generally has more than four carbons in them. When five carbons are bent around to a ring, the compound is cyclopentane (C_5H_{10}). Note that there are fewer hydrogens in cyclopentane than in normal or isopentane. Larger rings, starting with cyclohexane, are possible, too. This class of compounds is called naphthenes.

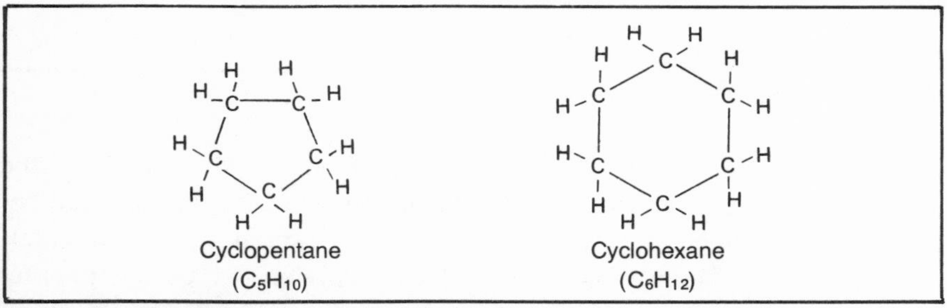

Fig. 5-5 —Naphthenes

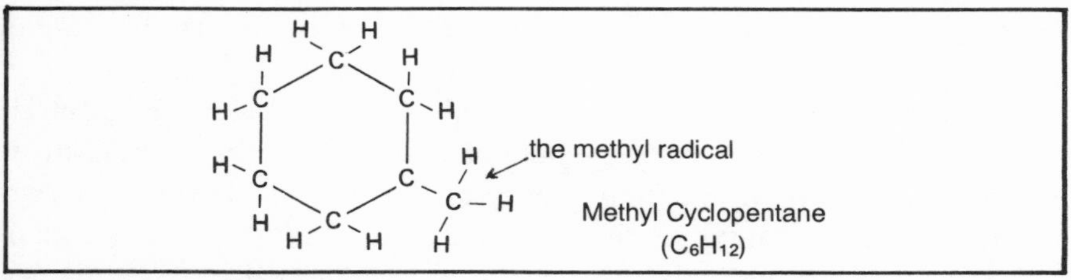

Fig. 5-6 —Methyl Cyclopentane

Beyond the simple paraffins and cyclic hydrocarbons are an infinite variety of possibilities by connecting the two types of compounds. The simplest example is methyl cyclopentane, which is the connection of CH_3, called a methyl radical, and a cyclopentane where a hydrogen ought to be, forming C_6H_{12}.

Olefins and Aromatics

It is possible to have a compound with two carbons and only four hydrogens. That would seem to violate the valence rules you just spent six minutes learning. But in the chemical compound ethylene, C_2H_4, there is a *double bond* between the two carbon atoms to make up for the deficiency of hydrogen atoms (Fig. 5-7). Ironically, the double bond holding the two carbons together is weaker than a single bond—not stronger. You could

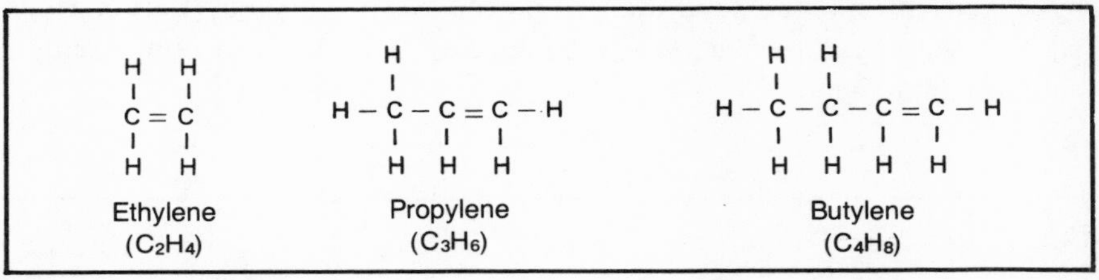

Ethylene (C_2H_4)   Propylene (C_3H_6)   Butylene (C_4H_8)

Fig. 5-7 —Olefins

think of it as two bonds sharing the spot of one bond. Therefore, the compound is unstable and can be chemically reacted with some other compound or element with relative ease to form a new compound, eliminating the double bond. That is why ethylene is such a popular compound for making more complicated compounds. For example, sticking a lot of ethylenes to each other makes polyethylene.

The key characteristic of the olefins is the absence of two hydrogens from an otherwise *saturated* paraffin i.e., a paraffin that has a full complement of hydrogens. The formula, therefore, is C_nH_{2n}.

Olefins are unnatural. It may be that only God can make a tree, but He could not handle olefins. They are not found in crude oil but are man-made in one of the several cracking processes that will be covered in later chapters. This seemingly irrelevant fact has implications in the design of refinery gas plants and will be discussed in Chapter VII.

The other olefins that will be of primary interest in petroleum refining will be propylene (C_3H_6) and butylene (C_4H_8). Like ethylene, these compounds can be reacted with other chemicals relatively easily and so are suitable for a number of both chemical and refinery processes.

Aromatics are another exception to the valence rules. Take the ring compound cyclohexane. Each carbon has two hydrogens attached and is attached to two other carbons. If you removed one hydrogen from each carbon, you could bring the valence rules back to being satisfied by putting in some double bonds between the carbons. A double bond between *every other* carbon will do it (Fig. 5-8). The resulting molecule is C_6H_6, *benzene*. Saying that every other bond around the benzene ring is a double bond is a gross simplification. The real description has to do with a resonating structure and hopping bonds.

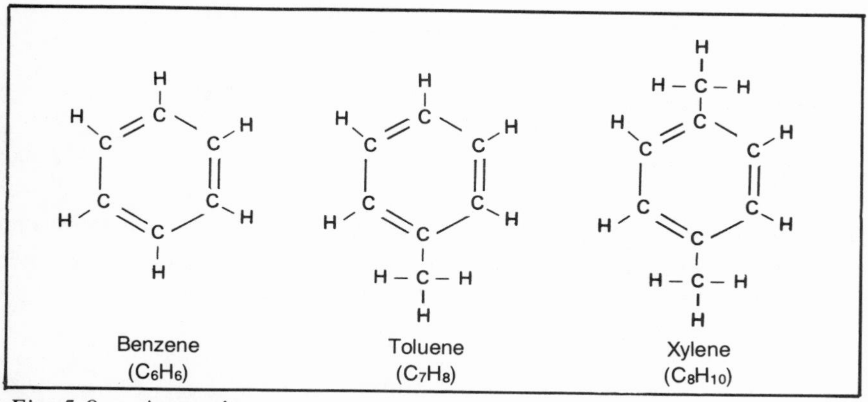

Fig. 5-8 —Aromatics

If one of the hydrogens of the benzene molecule is removed and CH_3 is put in its place, the result is C_7H_8, toluene. (The CH_3 is called a *methyl radical*. An ethyl radical would be C_2H_5. The radical names are similar to the paraffins they look like, methane and ethane.) If two methyl radicals are substituted for two, not one, hydrogens on the benzene ring, the result is C_8H_{10}, xylene (pronounced zi–leen).

The double bonds in the benzene ring are very unstable and chemically reactive. So benzene is a very popular building block in the chemical industry. Toluene and xylene are also important, and Chapter XVI deals with aromatics recovery in refineries.

The aromatics include *any compounds* that have the benzene ring in them. But in practice, many people use the term aromatics to refer to the *BTX's*, the benzene, toluene, and xylenes. The name aromatics, by the way, comes from the characteristic smell of most aromatic compounds, which is sort of sickly sweet hydrocarbonish odor.

You can see that once the number of carbons gets above six, the number of different variations in structure explodes. For that reason, very little attention is paid to the individual compounds. Sometimes, the proportion of paraffins, naphthenes, and aromatics might be used; usually the physical properties (gravity, viscosity, boiling temperatures, etc.) are used as an expedient.

EXERCISES

1. Isobutane is the isomer of normal butane. Why isn't there an isomer of propane?
2. How many different ways can isobutane be structured? Isopentane? Isobutylene?
3. Name the four types of structures generally referred to as the PONAs.
4. There are three different kinds of xylene. Can you draw the two not shown in this chapter?
5. Why is there only one type of toluene?

CAT CRACKING

In the adolescent years of the petroleum industry, the proportion of the crude oil barrel that consumers wanted in the form of gasoline increased faster than fuel oil. It became apparent to oil men that the production of enough straight-run gasoline to satisfy the market would result in a glut of heavy fuel oils. The economic symptoms of this problem were increasing prices of gasoline and declining prices of the heavier cuts.

To cope with this physical and economic problem, the inventive process engineers developed a number of cracking techniques, the most popular of which was *cat cracking*.

Process

In the last two chapters, the *chemistry* of cracking was discussed and the chemistry involved was explained. Here's the *process* of cracking: in a cat cracker, straight-run heavy gas oils are subjected to heat and pressure and are contacted with a *catalyst* to promote cracking.

A catalyst is a substance added to a chemical reaction which facilitates or causes the reaction; but when the reaction is complete the catalyst comes out just like it went in. In other words, the catalyst does not change chemically. It causes reactions between other chemicals. Catalysts are like some ten-year-olds you know. They never get into trouble; it's just that wherever they go, trouble happens.

The feed to the cat cracking process is usually a straight-run heavy gas oil as well as the tops from the flasher. The boiling point for cat feed can be anywhere in the 650-1100°F range. Heat is required to make the process go; temperatures in the cat cracker where the cracking takes place are about 900°F.

The process is designed to *promote* cracking in a specific way. *The object is to convert the heavy cuts to gasoline.* Ideally, all the product would be in the gasoline range, but the technology is not that good. During the cracking process, several phenomena occur. As the large molecules crack, the lack of sufficient hydrogen to saturate all the carbons causes some small amount of carbons to form coke, which is virtually pure carbon atoms stuck together. As the large molecules break up, a full range of smaller molecules from methane on up are formed. Due to the deficiency of hydrogen, many of the molecules are olefins. Where the large molecules in the feed are made up of several aromatic or naphthene rings stuck together, smaller aromatic or naphthenic compounds plus some olefins result. Finally, the large molecules, made up of several aromatic or naphthenic rings plus long side chains, are likely to crack where the side chains are attached. The resulting molecules, though lower in carbon count, are more dense or heavier; that is, their specific gravity is higher. They also tend to have higher boiling temperatures. The irony of this is that these molecules form a product which is heavier than the feed. The products of cat cracking are the full range of hydrocarbons, from methane through to the *residue range,* plus coke.

There are three parts to the cat cracking hardware: the reaction section, the regenerator, and the fractionator.

Reaction Section

The guts of the cat cracker is the reaction section (Fig. 6-1). The cat feed is pumped through a heater, mixed with catalyst being pumped into a line called a riser that goes to the bottom of a large vessel that looks like a water tank, called a reaction chamber. By the time the feed reaches the vessel, the cracking process is well underway, so the residence time in the vessel is only seconds. In fact, in the more modern cat crackers, all cracking takes place in the riser. The reactor is used only as a catalyst/hydrocarbon separator. That's done with a *cyclone,* a mechanical device using centrifugal motion.

The catalyst is generally one of two types: beads or particles. The beads are ⅛ to ¼ inch in diameter. The particles are much smaller and look like baby powder. Beads, however, are pretty much out of vogue now. The particles have two unusual characteristics. If a container of them is shaken or tilted, the powder reacts just like a fluid. Hence, the process is sometimes called *fluid cat cracking.*

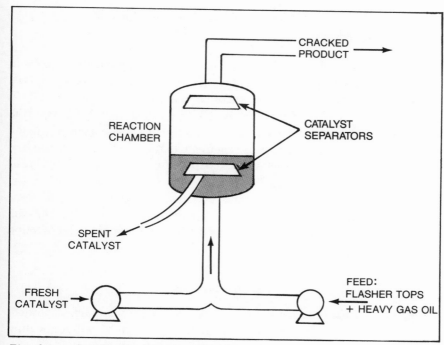

Fig. 6-1 —Cat Cracker Reaction Chamber

The second characteristic is common to both the beads and the particles but is not apparent to the naked eye. Under a microscope it can be seen that each has a large number of pores and, as a consequence, a tremendous *surface area*. Because the influence of the catalyst depends on contact with the cat feed, the tremendous surface area is important to the process.

The Regenerator

During the cracking process, the cracked hydrocarbon that is formed as coke ends up as a deposit on the catalyst, and as the catalyst surface is covered up, the catalyst becomes inactive (spent). To remove the carbon, the spent catalyst is pumped to a vessel called a regenerator (Fig. 6-2). Heated air, about 1,100°F, is mixed with the spent catalyst and a chemical reaction takes place:

$$C + O_2 \rightarrow CO \text{ and } CO_2 \text{ (older cat crackers)}$$
$$C + O_2 \rightarrow CO_2 \text{ (newer cat crackers)}$$

This process, *oxidation of coke,* is similar to burning coal or briquets in that carbon unites with oxygen and gives off carbon dioxide (CO_2), perhaps carbon monoxide (CO), and a large amount of heat. The heat, in the form of the hot CO/CO_2, is generally used in some other part of the process, such as raising the cat feed temperature in a heat exchanger. In the older cat crackers, the CO/CO_2 is sent to a CO furnace where oxidation of the rest of the CO to CO_2 is completed before the CO_2 is finally blown out to the atmosphere.

From the bottom of the regenerator comes freshly regenerated catalyst, ready to be mixed with cat feed on its way to the reaction chamber. Thus the catalyst is in continuous motion going through the cracking/regeneration cycle.

The Fractionator

Meanwhile, back on the hydrocarbon side, when the cracked product leaves the reaction chamber, it is charged (pumped in) to a fractionating

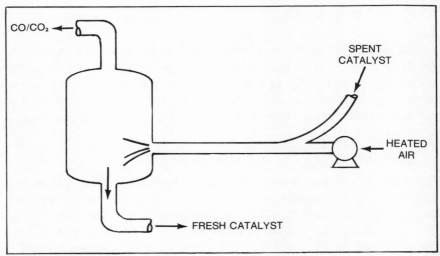

Fig. 6-2 —Catalyst Regenerator

column dedicated to the cat cracker product. The products separated generally are the gases (C₄ and lighter), cat cracked gasoline, cat cracked light gas oil, cat cracked heavy gas oil, and the fractionator bottoms, called cycle oil. A variety of things can be done with the cycle oil, but the most popular is to mix it with the fresh cat feed and run it through the reaction again. Some of the cycle oil cracks each time through the reactor. By recycling enough, all the cycle oil can be made to disappear. The process has the ominous designation *recycling to extinction*.

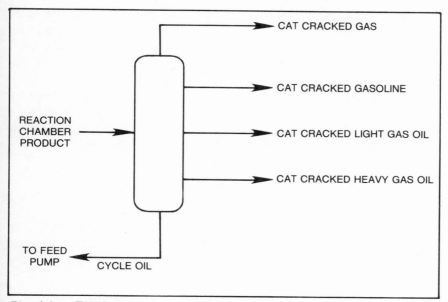

Fig. 6-3 —Fractionation

The cat cracked heavy gas oil can be used as feed to a thermal cracker or as a residual fuel component. The light gas oil makes a good blending stock for distillate fuel and the cat cracked gasoline a good motor gasoline blending component.

There is quite a bit of latitude in the cut point between the gasoline and light gas oil stream. This is one way to regulate the balance between gasoline and distillate as the seasons change. As the winter heating oil season comes on, many refineries go into a *max distillate mode*. One of the operations that changes is cat cracked gasoline end point, in order to drop more volume into the cat light gas oil. In the summer, during a *max gasoline mode*, the shift is in the other direction.

The light ends from the fractionator are different in composition than those from the crude distilling column light ends. The cracking process results in the creation of olefins, so the C4 and the lighter stream not only contain methane, ethane, propane, and butanes but also hydrogen, ethylene, propylene, and butylenes. Because of these extra components, this stream is sent to be separated at the *cracked gas plant*. This is in contrast to the gas from operations like distilling (and, as discussed later, the hydrotreater, hydrocracker, reformer, and others) where the gases contain only saturated compounds. These are sent to the *sats gas plant* for separation. The isobutane, propylene, and butylenes from the cat cracker will be of special interest when discussing alkylation, a process that converts these olefins into gasoline blending components.

All the pieces fit together as shown in Figure 6-4. You can see there are two circular flows going on. On the left side, the catalyst goes through the reaction, is regenerated and gets charged back to the reaction again. On the right side, hydrocarbon comes in and goes out, but the cycle oil causes continuous circulation of some components.

Yields

The object of the cat cracking process is to convert heavy gas oil to gasoline and lighter. A set of typical yields will demonstrate how successful the process is.

		% Volume
Feed:	Heavy Gas Oil	40.0
	Flasher Tops	60.0
	Cycle Oil	(10.0)*
		100.0
Yield:	Coke	8.0
	C4 and Lighter	35.0
	Cat Cracked Gasoline	55.0
	Cat Cracked Light Gas Oil	12.0
	Cat Cracked Heavy Gas Oil	8.0
	Cycle Oil	(10.0)*
		118.0

*Recycle stream not included in feed or yield total.

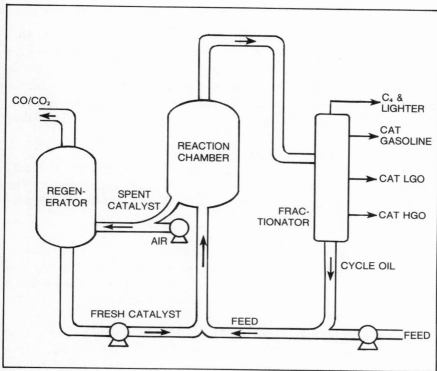

Fig. 6-4 —Cat Cracking Unit

Since the cycle oil is recycled to extinction, a simple diagram for cat cracking doesn't even show it as entering or leaving the process. Something more important does show up, however: the phenomenon of *gain*. In the yield structure shown above, the products coming out added up to 118% of the feed going in. That has nothing to do with the recycle stream but only with the gravities of the products coming out compared to the gravities of the feeds. If the yields were measured in percent weight instead of percent volume, the yields would come out to 100. But since most petroleum products are sold by the gallon, refiners measure everything in volume. And since cracking plays games with the densities, cat cracking yields show a substantial gain. Gain sometimes becomes an obsession with refiners as they attempt to "fluff up the barrel."

Process Variables

Usually a cat cracker is run to the limit of its coke-burning capacity. That can be reached in a number of ways, but it becomes apparent when the yields of gasoline fall off and either heavy gas oil or C_4 and lighter yields start to increase. Some of the things that affect cat cracker yields are the quality of the feed, the reactor temperatures, the feed rate, the recycle rate, and, incredibly enough, the time of day and temperature outside the control room.

Feed Quality. The cracking reaction is very complex, and lots of data are available to predict yields from different feed characteristics. The gravity of the feed and its paraffin, naphthene, and aromatic content are important.

Reactor Temperature. The higher the temperature, the more cracking will occur; but at some point, the amount of gases created will really take off, at the expense of the cat gasoline or cat light gas oil. The optimum reactor temperature is a function of economic incentive.

Feed Rate and Recycle Rate. The yields will suffer at higher feed rates, so the trade-off with the volume of fractionator bottoms being recycled or left in the cat heavy gas oil is watched.

Time of Day and Temperature. In order to regenerate the spent catalyst, fresh air is pumped into the regenerator continuously. As the temperature of the air outside goes down, the air gets more dense. Since the blowers that pump the air operate at fixed speed, more oxygen is actually pumped into the regenerator when it's cold than when it's hot. The more oxygen, the more coke burned off the catalyst. The fresher the catalyst, the better the reaction. The better the reaction, the more gasoline produced. From the operator's data logs, the swings can actually be plotted, and as night comes and the temperature drops, the yield gets better. With the heat of the afternoon, the yield falls off. Summer and winter have the same effects, which is too bad because usually the demands for gasoline go up in the summer when the yields go down.

Review   From a process point of view, the cat cracker can be sketched as a feed-in/product-out box in the refinery flow diagram. Thus far, you've covered distilling, flashing, and cat cracking. The way in which these come together is shown in Fig. 6-5. This diagram introduces the shorthand notation that will be used from here on.

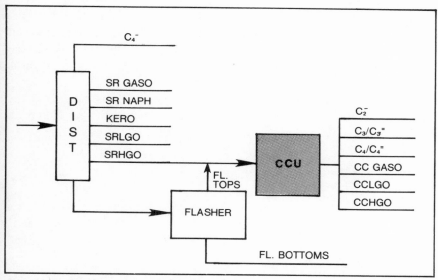

Fig. 6-5 —Refinery Flow Diagram

EXERCISES

1. Fill in the blanks:
   a. There are two circulating flows in a fluid cat cracker. On the one side is _____, on the other is _____.
   b. Spent catalyst has _____ deposited on it. The regeneration process removes it by reacting it with _____ to form _____ and _____.

c. The purpose of cat cracking is to convert _____ to _____.

d. Cat cracker feed usually comes from the _____ and the _____.

e. _____ are found in the cracked gas stream but not in the sats gas streams.

f. The cat-cracker fractionator bottoms, called _____ _____, are usually _____ _____ _____ by injecting them in the feed.

2. Use the data and answers from the previous chapters and the cat cracker yields from this chapter. Assume feeds to the cat cracker are all the flasher tops and all the straight-run heavy gas oil (the cut between the straight-run light gas oil and flasher feed.) How much cat light gas oil is produced?

3. An indicator of how complicated refining operations can get is the number of alternatives to make a downstream volume change. Name six ways to increase the volume of cat light gas oil.

REFINERY GAS PLANTS

*. . . and he shall separate them one from another, as a shepherd
divideth his sheep from the goats.*

—Matthew 25:32

Almost all the refinery processing units generate some volumes of
butane and lighter gas. Gas processing is not a very colorful operation, but
maybe that's why the engineers gave such exotic names to the parts.
There's *lean oil, fat oil, sponge oil, rectified absorbers,* to mention a few;
all are in the central location called the gas plant.

SATS GAS PLANT

The gas streams from most of the refinery process units have only
saturates in them: methane, ethane, propane, butanes, and maybe hydrogen. The word *saturates* is a synonym for hydrocarbons with no double
bonds; that is, all the carbon atoms are "saturated with hydrogen atoms."
These streams are usually handled in a facility called the *sats gas plant*.
This is in contrast to the streams that also contain the olefins (ethylene,
propylene, and butylenes) that are handled in the *cracked gas plant*.
Handling of hydrogen and hydrogen sulfide is so special that a special
chapter is devoted to it (Chapter XV).

The separation of the gases is a lot harder than the separations in the
crude distilling unit. Remember that each of those gases is a single
chemical and boils at a single temperature (Table 7-1). That doesn't leave
much tolerance for tail ends or any kind of sloppy separation. As a result,
the separation columns have lots of trays and lots of reboiling and
refluxing.

Another complication is the pressure/temperature relationships. The gas
plants are just the inverse of the flasher. In order to get part of the stream to
liquefy—an essential step in distillation—the mixture has to be either
supercooled (see the table below) or put under a lot of pressure. Usually
both methods are used in combination.

Some Properties of Light Ends

Compound	Formula	Boiling Temperature* (°F)	Density** (lbs/gallon)
Hydrogen	H_2	−423	—
Methane	C_1H_4	−258	2.5
Ethane	C_2H_6	−127	2.97
Ethylene	C_2H_4	−155	3.11
Propane	C_3H_8	− 44	4.23
Propylene	C_3H_6	− 54	4.37
Isobutane	C_4H_{10}	11	4.69
Normal Butane	C_4H_{10}	31	4.87
Isobutylene	C_4H_8	20	5.01
Normal Butylenes	C_4H_8***	21	5.01
	C_4H_8	34	5.09
	C_4H_8	39	5.23

*At atmospheric pressure
**At atmospheric pressure and 60 °F vacuum
***There are three forms of normal butylenes, each of them with the formula C_4H_8; each with
 a unique structure; each with unique properties

The sats gas coming from the processing units to the gas plant can go through the following typical steps:

1. *Compression.* Low pressure gases are compressed to about 200 psi.

2. *Phase separation.* Because of the high pressure, some of the gases liquefy, allowing them to be drawn off.

3. *Absorption.* The remaining high pressure gases are introduced to the

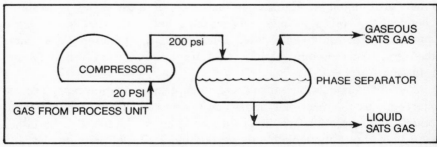

Fig. 7-1 —Compression and Phase Separation

middle of the *rectified absorber*. (Rectified is a synonym for distilling.) Into the top tray of the absorber is pumped a naphtha, which trickles down the column. As it does, it picks up (absorbs) most of the remaining propane and butane in the gas. The naphtha laden with the propane and butane is called *fat oil;* cleverly the naphtha that was introduced to the top of the column, ready to absorb the propane and butanes, is called *lean oil* (Fig. 7-2).

4. *Debutanizing*. The fat oil is charged to a column, called a debutanizer, which separates the butanes and propane from the naphtha. Because the boiling range of the naphtha starts about 180° and butane boils at about 32°F, the split is relatively easy (Fig. 7-3). It's certainly easier than splitting butane and propane from ethane in a distilling column. The liquid portion from the phase separator is charged to the debutanizer also.

5. *Depropanizing*. The C_3/C_4 stream is separated in a tall column that takes the propane overhead.

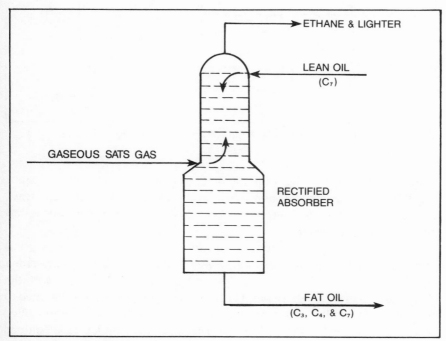

Fig. 7-2 —Absorption

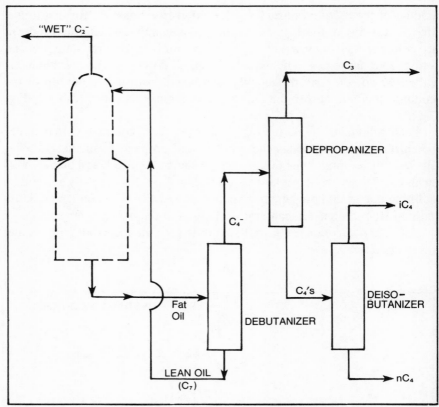

Fig. 7-3 —Debutanizing, Depropanizing, and Deisobutanizing

6. *Deisobutanizing.* Splitting the butanes is done in this column. Since the boiling temperatures of normal and isobutane are so close, many trays are needed to get good separation. The deisobutanizer is usually the tallest column in the gas plant.

7. *Sponge absorption.* One complication occurs when the lean oil is introduced to the rectified absorber. Some of the lean oil, usually heptane (C_7H_{16}) or octane (C_8H_{18}), vaporizes and goes out the overhead with the ethane and lighter (C_2-). To recover the lean oil, the overhead is charged to the bottom of another absorber column and another heavier lean oil, called *sponge oil*, charged to the top. The overhead is almost all C_2-, the bottoms almost all lean oil as shown in Fig. 7-4.

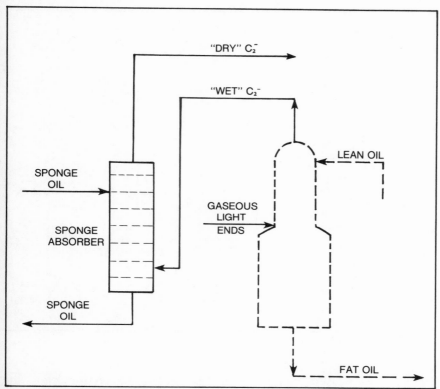

Fig. 7-4 —Sponge Absorption

The C_2- is usually not split in a refinery, since its primary use is refinery fuel. Sometimes, however, the ethane might be needed for feed to a chemical plant, and the C_2- is then de-ethanized in another absorber column.

All the steps are put together in Fig. 7-5.

Disposition

Why separate all these streams? Because each of the different components has found a unique home in the refining business, and the highest value can be attained only when the stream is separated for disposition to this use.

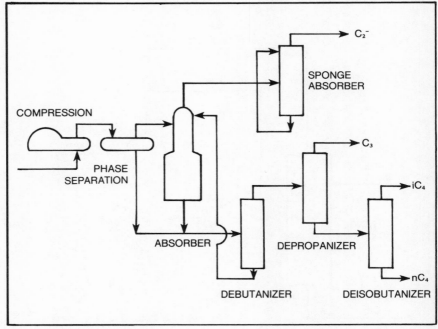

Fig. 7-5 —Sats Gas Plant

Isobutane.   Used almost entirely as one of the feedstocks to the alkylation process. Occasionally used as a motor gasoline blending component.

Normal butane.   Used almost entirely as a motor gasoline blending component. Due to its volatility, it is good for starting cold engines yet will stay dissolved in the gasoline with minimal evaporation. Some normal butane is also used as a chemical feedstock and as a fuel.

Propane.   In the U.S., most LPG (liquefied petroleum gas) is propane. About 10% of the LPG sales are butane or butane-propane mixes. Propane as a fuel has some unique characteristics that have found favor in many applications. It can be liquefied at reasonable temperatures and pressures for ease of transport, particularly in trucks. Yet at ambient temperatures, it will readily vaporize, making it easy to burn in stoves, home furnaces, etc. Propane has also widely been used as a chemical feedstock.

Ethane.   The only application that gives rise to separating ethane in a refinery is its use as a chemical feedstock. (See Chapter XVIII, "Ethylene Plants.") Otherwise, the ethane is left with the methane.

Methane.   Refineries have a voracious appetite for fuel. Methane finds a ready home in all the process furnaces and steam boiler furnaces around the refinery.

CRACKED GAS PLANT

The gas streams from the cat cracker and the other crackers contain olefins, ethylene, propylene, and butylenes. Usually the ethylene is not separated from the C2- stream going to the *refinery fuel system*. It could be, if there were a need in the chemical plant for it.

In a refinery, the propylene and butylenes are usually sent to an *alkylation plant*. For convenience, the paraffins, propane and butane are sent along as well and separated there (see Chapter VIII). Other options could include complete separation in the same fashion as in the sats gas plant.

Hydrogen

There are several processes in refineries that require hydrogen: hydro-cracking and a variety of hydrotreating units. Some of the gas streams, in particular the stream from the reformer, have higher concentrations of hydrogen. Often the gases from the reformer are kept segregated through the compression stage, and the compression permits separation of a *hydrogen concentrate,* an H_2/C_1 mix which can be used in the hydrogen applications.

Supplementary supply of hydrogen can be made specially if the refinery has a *hydrogen plant,* otherwise called a *steam-methane reformer,* which will be discussed in Chapter XV.

Storage Facilities

This is a good place to talk about the rather specialized facilities needed to store the refinery gases. Because of the volatility and boiling temperatures of these streams, high pressure containers are needed to store the streams in the liquid form.

Methane (and ethane not used as chemical feedstock) is usually never stored in a refinery but is sent directly to the fuel system as it is produced. There are some *surge drums* in the system which can accumulate some of the gaseous methane for short periods of time during operating changes. Also those tall *flares* so characteristic of refineries are used to take momentary surges in gas manufacture by burning the excess.

Propane and butanes (and sometimes ethane) can be stored in steel storage or in underground storage.

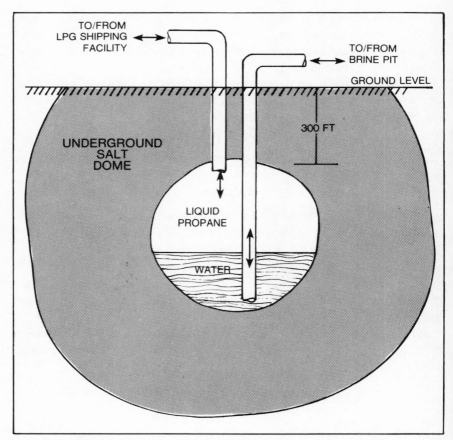

Fig. 7-6 —Underground Storage Jug for LPG

The characteristic of the propane and butane storage is the spherical shape. Steel storage is either in the form of *bullets* (cylinders on their sides with rounded ends) or large spheres. The round shape is due to an optimization of structural strength to accommodate the high pressure and the cost of the steel.

Underground storage generally takes one of two forms: *caverns* mined in rock, shale, or limestone; or *jugs* leached out of salt in underground *salt domes* as shown in Fig. 7–6.

In the former case, mined storage, the propane or butane is pumped in and out of the cavern in the liquid form, with the difference in the amount contained made up by more or less vapor being formed above the liquid.

In a salt dome, the ins and outs are handled in a more unusual fashion. The jug contains a combination of propane and salt water (brine). The two liquids, like any other oil/water combination, do not mix. To pump the propane out, more brine is pumped in, forcing the propane out. To fill the jug, propane is pumped in, forcing the brine out, usually to a brine pit.

One advantage of salt dome storage is its expandability. By substituting fresh water for the brine, more of the salt in the walls is dissolved, expanding the size of the jug at no cost. There is, of course, a practical limit to the size—the risk of collapse.

The other advantage is the original construction cost. Salt dome jugs are much cheaper than mined storage, which is much cheaper than steel storage. Unfortunately, Mother Nature limits the availability of salt domes, suitable rock, shale, and limestone to a few areas around the country.

EXERCISES

1. Fill in the blanks
   a. The two sections of the gas plant are called the _____ gas plant and the _____ gas plant.
   b. The naphtha used to strip out the heavier gases in the rectified absorber is called _____ _____. The liquids-laden lean oil is called _____ _____.
   c. A stream that can be used to absorb a lot of some parts of another stream is called a _____ _____.

2. Match up the following streams with the usual uses.

Methane	
Ethane	Hydrotreating
Propane	Blended to motor gasoline
Normal Butane	Refinery fuel
Isobutane	Alkylation feed
Propylene	Chemical Feed
Butylenes	Commercial fuel
Ethylene	
Hydrogen	

3. Draw a set of distillation columns that could be used to separate the cat cracker gases into C_2 and lighter, propane, propylene, normal butane, isobutane, and butylenes. Use Table 7-1 that has the boiling temperatures of each compound. Assume each column has only tops and bottoms, no side draws. What special problem does the butane/butylenes split present?

4. Make a flow diagram of the streams in and out of the units already covered, the crude distilling column, the flasher, the cat cracker, and the gas plant.

ALKYLATION

Yield who will to their separation,
My object in living is to unite.

—"Two Tramps in Mud Time," Robert Frost

After the engineers were so clever about the invention of cat cracking, they attacked the problem of all the light ends the process created. The objective was to maximize the volume of gasoline being produced, but butylenes and propylene were too volatile to stay dissolved in the gasoline blends. So a process was devised which was the inverse of cracking, *alkylation*, which starts with small molecules and ends up with large ones.

The Chemical Reaction

To a chemist, alkylation can cover a broad range of reactions; but to a refinery engineer, alkylation is the reaction of propylene or butylene with isobutane to form an iso-paraffin, called *alkylate* (Fig. 8-1).

The volumetric effect on refining operations is the inverse of cracking because there is a significant amount of shrinkage. With propylene as the feed, 1 bbl of propylene and 1.6 bbl of isobutane go in and 2.1 bbl of product comes out; 1 bbl of butylene and 1.2 bbl of isobutane yield 1.8 bbl of product. As in cracking, the weight in equals the weight out. Only the densities change.

The Process

The chemical reaction can be made to take place by subjecting the isobutane and olefins to very high pressures. But the equipment would be very expensive to handle this route to alkylation. Like a lot of other processes, *catalysts* have been developed to facilitate the process and simplify the hardware. For alkylation, the catalysts are usually either *sulfuric acid* or *hydrofluoric acid*. The processes using either are basically

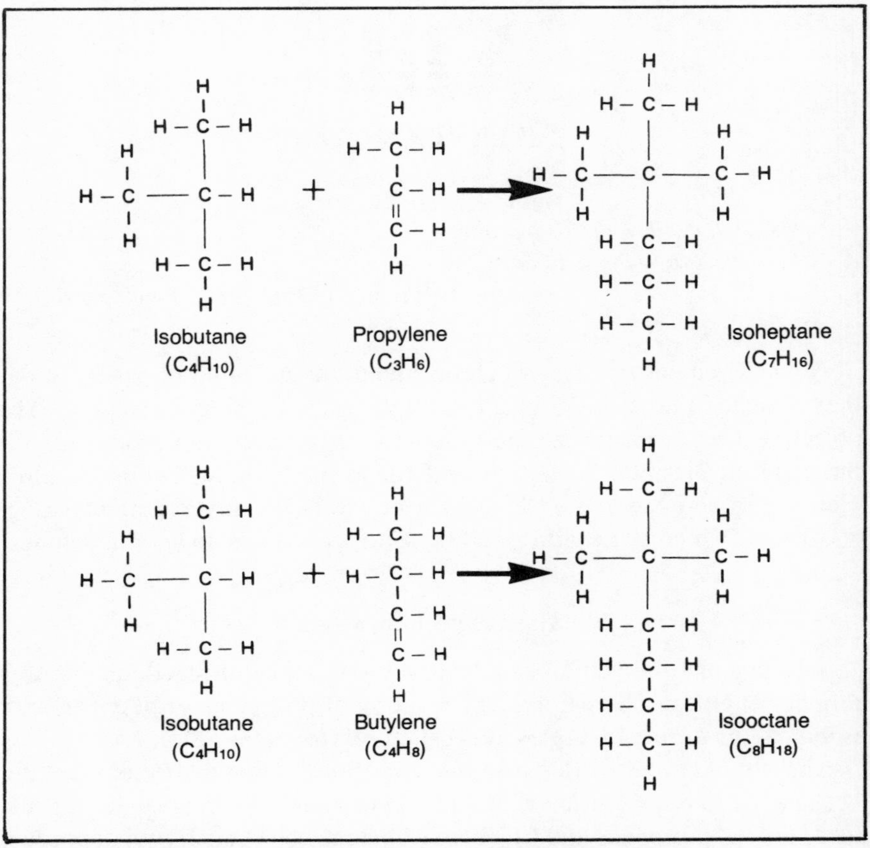

Fig. 8-1 —Alkylation of Propylene and Alkylation of Butylene

the same, but sulfuric acid seems to be much more popular, so only that will be covered here.

The *alky plant* consists of seven main parts: the chiller, the reactors, the acid separator, the caustic wash, and three distilling columns as shown in Figure 8-2.

The Chiller. Alkylation with sulfuric acid catalysts works best at temperatures in the neighborhood of about 40° F. So the olefin feed (a propane/propylene and/or butane/butylene stream from the cracked gas plant) is mixed with a stream of isobutane and a stream of sulfuric acid and

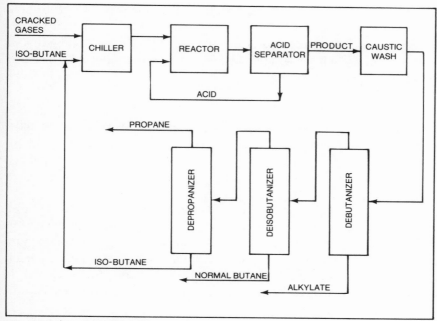

Fig. 8-2 —Alkylation Plant

pumped through a chiller. All the streams are under enough pressure to keep them in liquid form. Sometimes the chilling is done right in the reactor.

The Reactors.   The reaction time for the alkylation process is relatively slow, from 15 to 20 minutes, so the mixture is pumped into a battery of large reactors. The reactors hold so much total volume that by the time they all turn over once, the *residence time* of any one molecule is quite long (about 15 to 20 minutes). As the liquid passes through the reactors, it encounters mixers to assure that the olefins come in good contact with the isobutane in the presence of the acid, permitting the reaction to occur.

Acid Separator.   The mixture then moves to a vessel where no mixing takes place, and the acid and hydrocarbons separate like oil and water. The hydrocarbon is drawn off the top; the acid is drawn off the bottom. The acid is then recycled back to the feed side. The acid separator is also referred to as the *acid settler.*

Caustic Wash.   The hydrocarbon from the acid separator will have some traces of acid in it, so it is treated with *caustic soda* in a vessel. Caustic soda does to the hydrocarbon what Alka-Seltzer® does to your stomach when you have acid indigestion—it neutralizes the acid. The bad effects disappear. What's left is a mixture of hydrocarbons ready to be separated.

Fractionators.   Three standard distillation columns separate the alkylate and the saturated gas. The iso-butane is recycled to the feed.

Yields

During the alkylation process, a number of *side reactions* occur, some of which are more or less undesirable. Because there are a lot of molecules forming and reacting, there are small amounts of propane, butane, and pentane formed, which are not too bad; but a small amount of *tar,* a thick brownish oil containing complex hydrocarbon mixtures, forms as well. It usually settles out with the acid and disappears when the acid is sent back to the supplier for reprocessing.

The volumetric balances for propylene and butylene feed are as follows:

	Volume Balances	
Feeds	*Propylene*	*Butylene*
Propylene	1.0	—
Butylene	—	1.0
Isobutane	1.6	1.2
	2.6	2.2
Products		
Propane	0.3	
Normal Butane	—	0.1
Alkylate	1.8	1.7
	2.1	1.8

What the feeds and the yields don't show is the propane and normal butane that just pass through the plant untouched. As a matter of fact, the alky plant provides significant capacity to the refinery for making segregated propane and butane. When the alky plant is down for one reason or another, the propane/propylene stream is usually routed to the refinery fuel system and LPG production is cut significantly.

Process Variables

The alky plant manager has to watch a number of key variables to keep too many side reactions from occurring that could cause the quality of the alkylate to deteriorate, as evidenced by such things as lower octane number, poor color, and high vapor pressure.

Reaction temperature. Low temperatures cause the sulfuric acid to get syrupy and not to mix well. The olefins do not completely react. High temperatures cause compounds other than isoheptane and isooctane to occur, lowering the alkylate quality.

Acid strength. As the acid circulates through the process, it gets diluted with water that inevitably comes in with the olefins and it also picks up tar. As the acid concentration goes from 99% down to about 89%, it is drawn off and sent back to the acid supplier for refortification.

Isobutane concentration. By having an excess amount of isobutane, the process goes better. Isobutane recirculation systems are generally built in. The ratio of isobutane to olefin varies from 5:1 to 15:1.

Olefin space velocity. The amount of time the fresh olefin feed is resident in the reactor causes the quality of the alkylate to vary.

Review What's so important about aklylate? For one thing, it has a high octane number. For another, it has a low vapor pressure. Since gasoline blending has not been discussed yet, octane numbers and vapor pressure probably don't mean much to you. Suffice it to say that those are desirable qualities of gasoline blending components and that's the value of alkylation: low-valued cracked gas components are turned into high-valued gasoline components.

The alky plant can be represented by a box with propylene, butylene and isobutane *in,* along with propane and normal butane, and alkylate *out,* along with the propane and normal butane. To put alkylation in perspective, the refinery processing units covered so far, plus alkylation, are shown in Fig. 8-3. It's surprising how complicated it quickly gets.

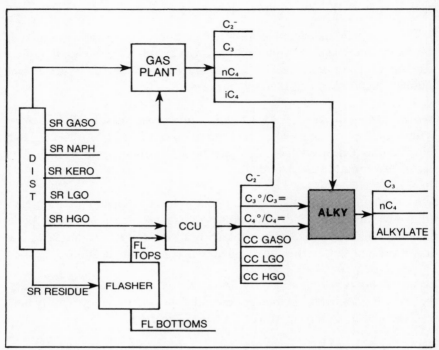

Fig. 8-3 —Refinery Processing Operations

EXERCISES

1. Fill in the blanks.
   a. Alkylation is the inverse of _____.
   b. The catalyst in alkylation is either _____ or _____,
      depending on the process chosen.
   c. The five main parts of an alky plant are:

d. Alkylate is made up mostly of two isoparaffins called _____ and_____.

e. Because of the high _____ _____ and low_____ _____, alkylate makes a good _____ _____ _____.

2. The cracked gas stream coming to an alky plant has the following composition.

Propane	.15
Propylene	.25
Isobutane	.10
Normal butane	.20
Butylenes	.30
	1.00

If the volume is 3000 B/D and the alky plant yields are those shown in this chapter, how much isobutane is needed from the sats gas plant? What are the alky plant outturns?

CATALYTIC REFORMING

Nothing so needs reforming as other people's habits.
—*Pudd'nhead Wilson*, Mark Twain

The third generation of refinery processes came as a result of the market requirement for increasing the quality of gasoline. Quality was generally measured by octane number which will be discussed in Chapter XII, "Gasoline Blending." In the 1950's and 60's the octane race was a prevalent factor in gasoline marketing. One method introduced to improve the octane number of gasoline was catalytic reforming. It is important for this discussion to keep a mental finger in Chapter V, "The Chemistry of Petroleum."

The Chemical Reactions

Unlike the previous processes discussed, there is very little difference in the boiling range of the feed and the product of a catalytic reformer. What does change is the *chemical* composition.

Feed to a cat reformer is usually *straight run naphtha*. Naphthas for some other processes, such as thermal cracking, coking, and especially hydrocracking, are sometimes fed to the cat reformer also. Typically there is a high concentration of paraffins and naphthenes in this fraction. The cat reformer causes many of these components to be reformed into aromatics compounds, which have much higher octane numbers.

Cat reforming typically results in the following change in the naphtha:

	% Volume	
	Feed	Product
Paraffins	50	35
Olefins	0	0
Naphthenes	40	10
Aromatics	10	55

The good reactions that take place in the cat reforming process are mainly:

1. Paraffins are converted to isoparaffins.
2. Paraffins are converted to naphthenes.
3. Naphthenes are converted to aromatics.

And some not-so-good reactions take place:

4. Some of the paraffins and naphthenes crack and form butanes and lighter gases.
5. Some of the side chains get broken off the naphthenes and aromatics and form butanes and lighter gases.

The important thing to remember is *paraffins and naphthenes get converted to aromatic compounds and some isomers,* as shown in the reactions in Fig. 9-1.

The Hardware

You might expect some unusual hardware would be required to cause these complicated reactions to take place. On the contrary, what's needed is an *unusual catalyst,* and in this case it's made of alumina, silica, and platinum. And the platinum is in no small amounts (several million dollars' worth in one process unit), so great care is taken to keep track of it.

There are several ways of putting the hydrocarbon in contact with the catalyst. The one covered here is called *fixed bed* because the hydrocarbon is dribbled through the catalyst, which stays put in a vessel.

The operating conditions which promote each of the chemical reactions in Fig. 9-1 are different, as measured by pressure, temperature, and residence time. For that reason, three reactors in series (Fig. 9-2) are used, each one doing a different job. The reactors operate at 200–500 psi pressure and 900-975°F. The vessels are characteristically spherical in shape.

The naphtha feed is pressurized, heated and charged to the first reactor, where it trickles through the catalyst and out the bottom of the reactor. This process is repeated twice in the next two reactors. The product is then run through a cooler where much of it is liquefied. The purpose of the liquefication at this point is to permit separation of the hydrogen rich gas stream for recycling. This process is important enough to warrant a few sentences.

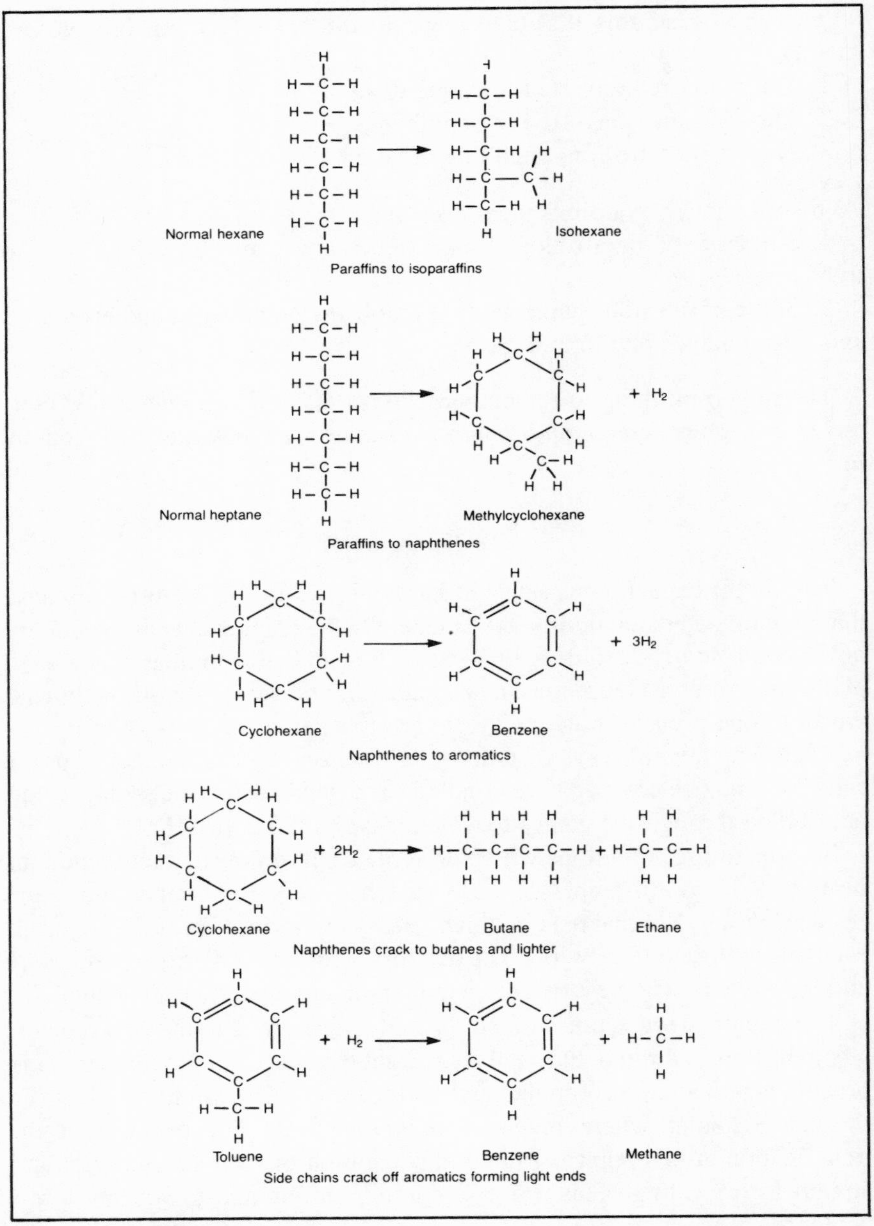

Fig. 9-1 —Reformer Reactions

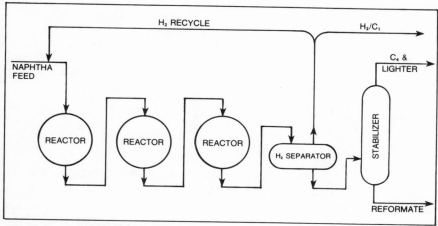

Fig. 9-2 —Catalytic Reforming

Hydrogen is an important by-product of cat reforming. Look back at the chemical reactions. Most of them result in the production of extra hydrogen because aromatics don't need as many hydrogens as naphthenes and paraffins. But reforming is also a user of hydrogen in the reactors. Hydrogen must be mixed with the feed to keep a high concentration of hydrogen vapor in the reactors. This prevents carbon atoms from depositing on the catalyst, as in cat cracking. Instead, the carbon reacts with the hydrogen and forms a hydrocarbon gas.

Meanwhile back at the hardware, part of the hydrogen stream is recycled to the feed while the other part is sent to the gas plant. The liquid product from the bottom of the separator is sent to a fractionator called a *stabilizer* which is nothing more than a debutanizer. It makes a bottom product called *reformate;* butanes and lighter go overhead and are sent to the sats gas plant.

Regeneration

After an interval of operations, coke deposits on the catalyst cause a decline in activity. Symptoms of the decline are reduced octane numbers, lower reformate yield per barrel of feed, or both, as shown below.

Reformers of an early vintage were shut down to regenerate the catalyst, but a continuous flow design was devised by adding another reactor. At any

	Yields-% Volume	
	Fresh or Regenerated Catalyst	Aged Catalyst
H_2	2	2
C_1/C_2	2	3
C_3	2	3
iC_4	3	4
nC_4	3	4
Reformate	88	84
(Octane number)	(94)	(92)

one time three of the reactors are in operation, the fourth reactor being in the regeneration mode. Regeneration is accomplished by admitting hot air to remove the carbon from the catalyst, forming carbon monoxide and dioxide. The cycle time for a reactor to be off line is only about 30 hours, so the catalyst is kept fresh virtually all the time.

Despite the continuous regeneration, over a long period of time the activity of the catalyst will decay. The high temperatures required for regeneration cause the catalyst's pores to collapse. Consequently, every 2-3 years the entire reformer must be shut down for catalyst change out.

Process Variables

The dials and buttons on a reformer that the engineers play with are the temperature, pressure, and residence time. The game is really a balance between the volume of the reformate and its quality. Fig. 9-3 shows the relationship: *as the octane number goes up, the percent volume reformate goes down.* Correspondingly, the yield of butanes and lighter goes up. So the operation of the cat reformer must be tuned in very closely with the gasoline blending operations and the gasoline component yields of the other processing units.

The properties of the naphtha feed, as measured by the paraffin, olefin, naphthene, and aromatic content, affect the yields and quality also. The PONA analysis of naphtha is an important component in the analysis of a crude oil's value.

Aromatics.   Cat reforming is a primary source for benzene, toluene, and xylenes. The process for recovery of the BTX's will be covered separately in a later chapter.

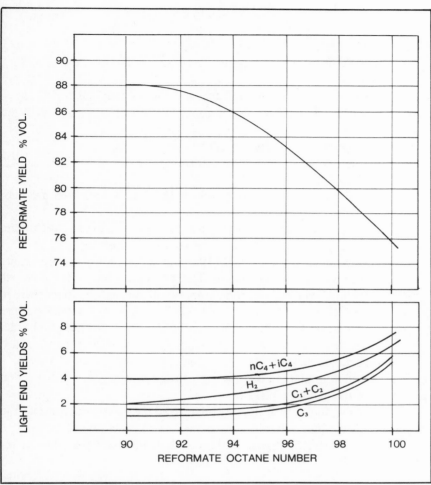

Fig. 9-3 —Cat Reformer Yields vs. Octane Number

Review.   Catalytic reforming is an important process for upgrading low octane naphthas to a high octane blending component, reformate. The PONA content of the feed is shifted from the P and N toward the A to capitalize on the high octane numbers of aromatics. Unfortunately, the higher the octane number of the reformate, the lower the yield and the more light ends produced.

EXERCISES

1. Fill in the blanks:
   a. The basic purpose of catalytic reforming is to increase the _____ of naphtha.
   b. The expensive component of the reforming catalyst is _____.
   c. Reformate has a much higher concentration of _____ than the reformer feed.
   d. An important by-product of cat reforming is _____.
   e. The other by-products of cat reforming are _____, _____, _____, and _____.
   f. To keep the catalyst from getting coked up, the reactors are kept filled with _____.
   g. Symptoms of aging catalyst are the decline of the reformate _____ or _____, and the increase yield of the _____.

2. Suppose the reformer feed is 15 M B/D and the operating conditions are set to produce 91 octane reformate. The 91 octane reformate is worth 100¢/gal and the C4 and lighter streams are worth 50¢/gal. Each increase in octane number increases the reformate value by 1¢/gal. Does it make sense to increase the severity of the reformer, that is, run it to produce higher octane reformate? Use the yield versus octane number chart in Fig. 9-3. (Hint: check the economics at 91, 95, and 100 octane).

3. Draw an additional example of each of the five types of reactions that take place in the catalytic reformer.

4. In the 1960's, Shell Oil Company advertised the use of Platformate® in its premium gasoline. What is Platformate® a contraction of?

5. Draw the refinery configuration covered so far, including the cat reformer in its place.

RESIDUE REDUCTION

The very ruins have been destroyed.

—Civil War, Luca

High rates of crude runs inevitably produce large volumes of the *bottom of the barrel*. In the U.S. economy, the change in the demand for gasoline has never been matched by the change for residual fuel. Even the reduction of straight-run residue volumes by flashing could not balance residual fuel supplies with demands, so refiners resorted to several types of process units to convert residue to light products. As early as 1920, large volumes of pitch were being processed in *thermal crackers,* greatly ameliorating the gasoline/residual imbalance. In recent years, advanced technology has favored construction of *cokers*. These processes are very similar to each other and will be covered here as the most popular methods of residue reduction.

It's important to note that there is sometimes a fuzzy understanding of the difference between *residue reduction* and *pitch destruction*. The former does not always imply the latter. There will be more on this subject at the end of the chapter.

Thermal Cracking

In the chapter on vacuum flashing, you learned about a process unit designed to *avoid cracking* of molecules due to high temperature. In this chapter, you will learn about *promoting cracking* due to high temperatures. The difference is in the controlled conditions.

Remember, thermal cracking is the breaking of hydrocarbon molecules into smaller compounds, usually olefinic due to the absence of extra hydrogen. Long chain paraffins can rupture anywhere. For the cyclic compounds, the break tends to be at the point where a straight chain, if any, is attached. As a consequence, the heavier products of cracking tend to have high olefin, naphthene, and aromatic contents.

Feed to the thermal cracker is usually flasher bottoms (pitch), but sometimes cat cracked heavy gas oil and cat cracked cycle oil are used for feed.

If the broader range of streams is fed to the thermal cracker, the lighter distillate range hydrocarbons are kept separate from the heavier stocks. Though they are shown combined in Fig. 10-1, each is fed to a separate furnace, since the temperature requirements (severity) are higher for the lighter products. The furnaces heat the feed to the 950-1020°F range. Residence time in the furnaces is kept short to prevent much reaction from taking place in the tubes going through the furnace. Otherwise, the formation of coke can take place, quickly clogging the furnace tubes, shutting the operation down. The heated feed is then charged to a *reaction chamber* which is kept at a pressure high enough (about 140 psi) to permit cracking but not coking.

From the reaction chamber the product is mixed with a somewhat cooler *recycle stream*, which stops the cracking. Both streams are charged to a *flasher chamber*, where the lighter products go overhead because the pressure is reduced as in a vacuum flasher. The bottoms are a heavy

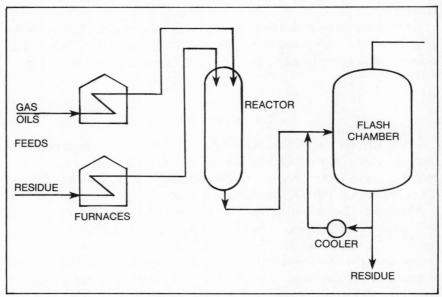

Fig. 10-1 —Thermal Cracker Reaction Section

residue, part of which is the *recycle stream* for the reaction chamber; the balance is usually blended into *residual fuel*.

The lighter products from the top of the flash chamber are charged to the *fractionator*, as shown in Fig. 10-2. The C4 and lighter streams are sent to the *cracked gas plant*. The thermal cracked gasoline and naphtha are used for gasoline blending or can be sent to the reformer. The gas oils can be used as a distillate fuel or, like the fractionator bottoms, can be recycled to extinction.

Coking

Coking is severe thermal cracking. Over the years, as more knowledge was learned about thermal cracking, it was found that high temperatures and very high velocities postpone the formation of coke until the pitch gets from the furnace tubes to a large surge tank. Retaining the hydrocarbon in this insulated vessel called a *coke drum* permits extensive and controlled cracking and coking. Technology has been developed to handle this process efficiently on a continuous flow basis. While the concept was simpler

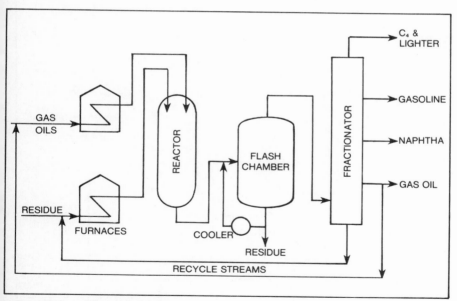

Fig. 10-2 —Thermal Cracker

than thermal cracking, the hardware was mechanically more complicated because of the *coke handling* facilities. Since coke is a solid, unique problems were presented.

The feed to the coker (same ones as a thermal cracker) is heated to about 1000°F and then charged to the bottom of a coke drum. The cracked lighter product rises to the top of the drum and is drawn off. The heavier product remains and, because of the retained heat, cracks to coke, a solid coal-like substance. Vapors from the top of the drum are sent to a fractionator for separation, just like the thermal cracker.

Coke removal from the drum is a special problem because it is a solid cake. Back in the old days, the thermal cracker reaction chamber sometimes got *coked up* because of some upset or accident. The only way to get the coke out was to send workers into the vessel with chipping hammers and oxygen masks. Surely this is what inhibited the development of coke production in refineries.

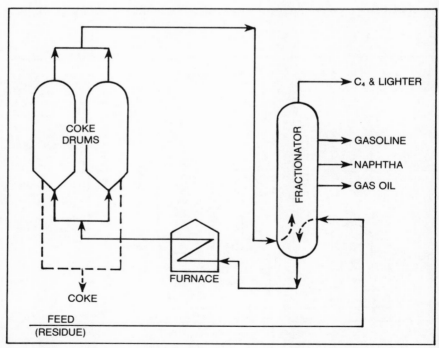

Fig. 10-3 —Coker

Nowadays, *decoking* is a routine daily occurrence and is usually accomplished by using a high pressure water jet (about 2000 psi). First a hole is drilled in the coke from the top to the bottom of the drum. Then a rotating stem is lowered through the hole, spraying a water jet sideways. The high pressure cuts up the coke into lumps, and it is dropped out of the bottom of the drum into trucks or rail cars for shipment or hauling to the coke barn.

Typically the drums are on about a 48 hour cycle; filling a drum with coke takes about 24 hours. Switching, cooling, decoking, and emptying take about 22 hours, during which time another drum is filling.

Yields. The outturns from thermal cracking and coking are sensitive to operating conditions, mainly furnace temperature, and properties of the feed. Boiling range, PONA analysis, and gravity are usually used to predict yields. As an example of what these units can do, feeding flasher bottoms from a West Texas crude to a thermal cracker being run to maximize gasoline or to a coker might operate as follows:

	Thermal Cracker Yields	Coke Yields
Pitch Feed	1.00	1.00
Products		
Coke*		.30
C₂ and lighter	.18	.15
Propane	.05	.01
Propylene	.07	.01
Butanes	.04	.02
Butylenes	.06	.04
Gasoline	0.30	0.08
Naphtha	0.15	0.15
Gas Oil	0.10	0.50
Residue	0.20	—

*Percent weight, since it's not liquid.

The thermal cracker makes about 80% of the residue go away. The coker makes it all go away, but it also makes 30% coke for which a market must be found. Both processes also make gasoline, naphthas, and gas oils that are low in quality. The naphtha is often processed further in a cat reformer and the gas oil in a hydrocracker. But the residue goes away, and that's the name of that game!

Coke. A few words about coke might be helpful. Most coke produced in refineries is sponge-like in appearance; hence the name *sponge coke*.

The main uses for sponge coke are the manufacture of electrodes and anodes, a carbon source for carbides, and the manufacture of graphite. The strength of sponge coke is not sufficient for its use in blast furnaces for pig iron or in foundry work.

A second form, *needle coke*, derives its name from its microscopic elongated crystalline structure. Needle coke requires special coker feeds and more severe operating conditions. Its qualities make it preferred to sponge coke in the manufacture of electrodes. All these factors combine to increase its value well above sponge coke.

Coke usually has water and some liquid hydrocarbon mixed in it, and before it is used, it must be dried. Crushing and heating coke to drive off this material is called *calcining*.

Coke can also be used as refinery fuel in the same way as another solid, coal.

Visbreaking

A visbreaker is a poor-man's thermal cracker. The visbreaker takes pitch (flasher bottoms) and thermally cracks part of it. That permits reflashing the pitch to reduce the volume. But then some diluent, called *cutter stock*, has to be added to the visbreaker pitch to make it loose enough to be marketable as residual fuel. The benefit of the scheme is that the cutter stock added back is less than the cracked product freed up by visbreaking, so there is a net reduction of residual fuel being produced.

The cutter stock can be cat cracked heavy gas oil, cycle oil, or even a cut of the flashed distillate from the visbreaker.

Visbreaking sounds a lot like thermal cracking but differs in intensity. The hardware is simpler and cheaper to operate. On the other hand, only about 20–30% of the pitch is destroyed.

Comment   In many parts of the world outside the U.S., a significant amount of straight run heavy gas oil is left in the residue. In the U.S., it would have been segregated as flasher tops and replaced using extra heavy cracked gas oils as cutter stock.

Residue reduction in the context of these sources of residual fuel could be flashing/cat cracking as opposed to thermal cracking/coking. In such case, the process is residue reduction but not pitch destruction.

EXERCISES

1. Fill in the blanks.
   a. The main difference between thermal cracking and cat cracking is that the former uses no _____ to promote cracking.
   b. The main difference between thermal cracking and coking is the production of _____ and _____.
   c. To prevent coking in a thermal cracker reaction chamber, the feed is _____.
   d. To promote coking in a coke drum, the feed is _____.
   e. Thermal cracker and coker gasoline, naphthas, and gas oil are _____ in quality.
   f. Two types of refinery coke are _____ and _____.
   g. The C4 and lighter gas is sent to the _____ gas plant because it contains _____.
   h. The three main parts of thermal crackers or cokers are the _____, _____, and the _____, which in a coker is called a _____ _____.

2. The refinery manager in chapter two and three has to tell the sales manager how much coke to sell. He runs all the flasher bottoms to the coker. Use the yields in this chapter and answers to problem 2 in chapter three to do the calculation. Assume the gravity of the flasher bottoms is 10° API or 350 lb/bbl. (Remember—the coke yield is percent weight, and the units of sale for coke are short tons, 2000 lb per ton).

3. Add a thermal cracker to your drawing of the refinery process units and stream flows.

XI.

HYDROCRACKING

Cancel and tear to pieces that great bond.

—*Macbeth*, Shakespeare

Hydrocracking is a later generation process than cat cracking or cat reforming, but it was designed to accomplish more of what each of those processes do. Hydrocracking can increase the yield of gasoline components, usually at the expense of gas oil range process stocks. It produces gasoline blending components of a quality not obtainable by recycling the gas oils through the cracking process that generated them. Hydrocracking can also be used to produce light distillates (jet fuel and diesel fuel) from heavy gas oils. Maybe best of all, however, is that hydrocracking produces no bottom-of-the-barrel leftovers (coke, pitch, or resid). The outturn is all light oils.

The Process

Hydrocracking is simple. It's cat cracking in the presence of hydrogen. The combination of the hydrogen, the catalyst, and the operating conditions permit cracking low quality light gas oils from the other crackers that would otherwise be made into distillate fuel. The process gives a high yield of good quality gasoline.

Think for a moment about the implications of having a hydrocracker. Its single most important advantage is its ability to *swing the refinery outturns* from high gasoline yields (when the hydrocracker is running) to high distillate yields (when it is shut down).

Hydrocracking operations are a lot like that old, cynically disparaging remark of the coach who just lost a player to a rival team: "It probably improves the quality of both teams." Hydrocracking improves the quality of both the gasoline blending pool and the distillate pool. It intakes the worst of the distillate stocks and outturns a better-than-average gasoline blending component.

There is one other thing worth noting: Hydrocracking produces a relatively large amount of isobutane, which is useful in balancing feed to the alky plant.

About a dozen different types of hydrocrackers are presently popular, but they all are very similar to the one described below.

The Hardware and the Reactions

Hydrocracking catalysts are fortunately less precious and expensive than reforming catalysts. Usually they are compounds of sulfur with cobalt, molybdenum, or nickel plus alumina. (You probably always wondered what anyone used those metals for.) In contrast to cat cracking but like cat reforming, hydrocrackers have their catalysts in a fixed bed. Like cat reforming, the process is carried out in more than one reactor, in this case two, as shown in Fig.11-1.

Feed is mixed with hydrogen vapor, heated to 550-750°F, pressurized to 1200-2000 psi, and charged to the *first stage reactor*. As it passes through the catalyst, about 40 to 50% of the feed is cracked to gasoline range material (below 400°F end point).

The hydrogen and the catalyst are complementary in several ways. First, the catalyst causes cracking, but the cracking process needs heat to keep it going. That is, it is an *endothermic process*. On the other hand, as the cracking takes place, the hydrogen saturates (fills out) the molecules, a process that gives off heat. This process, called *hydrogenation*, is *exothermic*. Thus, hydrogenation gives off the heat necessary to keep the cracking going.

Another way in which they are complementary is in the formation of isoparaffins. Cracking forms olefins, which could join together to form normal paraffins. Hydrogenation rapidly fills out all the double bonds, often forming isoparaffins, preventing reversion to less desirable molecules. (Isoparaffins have higher octane numbers than normal paraffins.)

After the hydrocarbon leaves the first stage, it is cooled and liquefied and run through a hydrogen separator. The hydrogen is recycled to the feed. The liquid is charged to a fractionator. Depending on the products desired (gasoline components, jet fuel, and gas oil), the fractionator is run to cut out some portion of the first stage reactor outturn. Kerosene range material could be taken as a separate side draw product or could be included in the fractionator bottoms with the gas oil.

The fractionator bottoms are again mixed with a hydrogen stream and

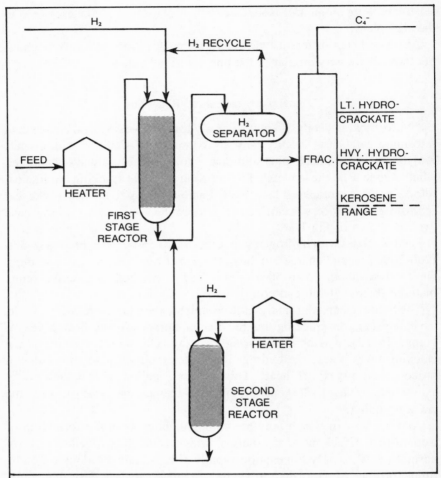

Fig. 11-1 —Two Stage Hydrocracker

charged to the second stage. Since this material has already been subjected to some hydrogenation, cracking, and reforming in the first stage, the operations of the second stage are more severe (higher temperatures and pressures). Like the outturn of the first stage, the second stage product is separated from the hydrogen and charged to the fractionator.

Imagine the kind of vessels necessary to handle the operating conditions, 2000 psi and 750°F. The steel reactor walls are sometimes 6 in. thick. A critical worry is the possibility of *runaway cracking*. Since the overall process is endothermic, rapid temperature increases are possible, accelerating the cracking rates dangerously. Elaborate quench systems are built into most hydrocrackers to control runaway.

Yields. Another excellent feature of the hydrocracker is a 25% gain in volume. The cracking/hydrogenation combination results in products whose average gravity is a lot higher than the feed. A typical set of hydrocracker yields, using coker gas oil and cat cracker light oil as feed, are shown below. The primary products, gasoline blending components, are called hydrocrackate.

	Volume Balance
Feed	
Coker gas oil	0.60
Cat cracked light gas oil	0.40
	1.00
Product	
Propane	
Iso-butane	0.02
Normal butane	0.08
Light hydrocrackate	0.21
Heavy hydrocrackate	0.73
Kerosene range	0.17
	1.21

Not shown is the hydrogen requirement, which is measured in standard cubic feet per barrel of feed. Net consumption of 2,500 scf/bbl is typical.

Heavy hydrocrackate is a naphtha that contains a lot of aromatics *precursors* (compounds that easily convert to aromatics). It is often fed to the reformer for upgrading. The kerosene range material makes a good jet fuel or a distillate blending stock because of its low aromatic content, a consequence of hydrogen saturating those double bonds. For more information, see Chapter XIII, "Diesel Fuels," and Chapter XIV, "Jet Fuel Hydrotreating."

Residue Hydrocracking. There are a few hydrocrackers that have been constructed to handle straight run residue or flasher bottoms as feed. Most of them are operated as hydrotreaters, as described in Chapter XV. The yields are over 90% residual fuel. The purpose of the operation is to remove sulfur by the catalytic reaction of the hydrogen and the sulfur compounds, forming H_2S, hydrogen sulfide. Residue with a sulfur content of about 4% or less can be converted to heavy fuel oil with less than 0.3% sulfur.

Review   With the addition of the hydrocracker to the refinery processing scheme, the absolute requirement for integrated operations becomes apparent. In one sense, the hydrocracker is the pivotal unit since it can swing the refinery yields between gasoline, distillate fuel, and jet fuel. But just as important are the feed rates and operating conditions of the cat cracker and the coker. In addition, the alky plant and reformer are necessary appendages for economical disposition of the cracker products.

EXERCISES

1. Contrast hydrocracking, cat cracking, and thermal cracking in terms of feed, what promotes the reaction, and what the products and product PONA's are.
2. How is a hydrocracker complementary to a cat cracker? A reformer to a hydrocracker?
3. Draw the refinery flow diagram with the hydrocracker added.

XII.

GASOLINE BLENDING

The good things in life are not to be had singly, but come to us with a mixture.

—Charles Lamb

The best-known hydrocarbon is gasoline, but it is astounding how little is known about the characteristics of this product. The explanation for this public naivete probably lies in the competitive pressures which force producers to make a product which performs. And if the product performs, the buying public loses interest (or never has any) in what makes it good.

This chapter will cover several areas:

1. The two most important variables in gasoline blending: vapor pressure and octane number.
2. The effect of tetraethyl lead in gasoline.
3. Procedures for blending gasoline.
4. The impact of gasoline blending requirements on refinery operations.

There is a small chance that you'll be lost from the beginning if you don't have some idea how an automobile engine works, so a couple of paragraphs and illustrations will precede the discussion.

Gasoline Engines.   The essential parts of a gasoline engine, at least for this discussion, are the gas tank, fuel pump, carburetor, cylinder, piston, and the spark plug. Engines without the last item (spark plugs) will be discussed in the next chapter because they're diesels.

You might say the process flow in a gasoline engine starts in the gas tank when you fill it at a gas station. Then when you start the engine, the fuel pump sucks gasoline out of the tank and sends it to the carburetor. The carburetor vaporizes the gasoline, mixes it with air, and sends it to the cylinder. At this point, the *sequence* of events illustrated in Fig. 12-1 is essential.

The gasoline/air vapor mixture is sucked into the cylinder as the piston moves towards the bottom of the stroke and the space in the cylinder grows to its largest. The inlet valve closes and the piston then moves up the cylinder, *compressing* the vapor. When the piston reaches the top of the stroke, the spark plug gives off a powerful spark, igniting the gasoline. The gasoline burns explosively, causing a huge expansion of the gases and pressure on the piston. Power is transmitted to the crank shaft as the piston is forced down the cylinder in a *power stroke*. At the bottom of the power stroke, the exhaust valve at the top of the cylinder opens and the burnt fuel is pushed out as the piston moves up the cylinder. At the top of the stroke, the inlet valve opens again and the process is ready to be repeated. Note that each cycle requires two trips of the piston up and down the cylinder.

Vapor Pressure

One of the crucial steps in the engine cycle is the vaporization of the gasoline. When the engine is warmed up, there is no problem: the engine heat assures that 100% of the gasoline will enter the cylinder in vapor form. But when the engine is being started from cold, the conditions are much more difficult.

The trick in handling cold starts is to have enough volatile hydrocarbon in the gasoline to get a vapor-air mixture that will ignite. The measure of volatility is vapor pressure, and more specifically *Reid Vapor Pressure* (RVP), named after the man who designed the test apparatus.

Definition. Vapor pressure is a measure of the surface pressure it takes to keep a liquid from vaporizing. A light hydrocarbon like propane will have a very high vapor pressure, since it is very volatile. A heavier hydrocarbon like gas oil will have nearly a zero vapor pressure, since it will vaporize very slowly . . . at normal temperatures. If you think for a moment you'll realize that vapor pressure is a function of temperature. RVP is measured at 60°F.

Engine Conditions. Definitions out of the way, go back to the carburetor problem. The RVP of gasoline must meet two extreme conditions. On cold starts, enough gasoline (maybe 10%) must vaporize to provide an ignitable mixture. Once ignition occurs, the rest of the gasoline that didn't vaporize will burn too. The other extreme is when the engine is running while completely warmed up or, even more extreme, is being restarted when it's hot. At that point, the gasoline vapor must not expand so much

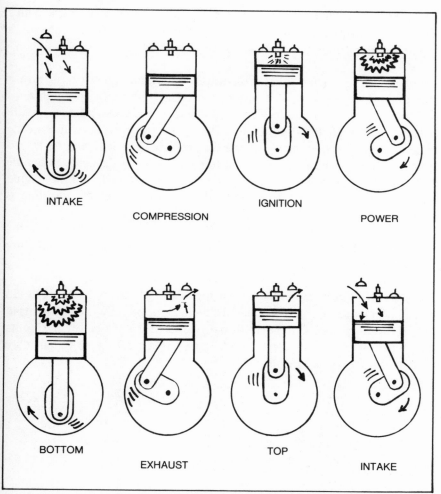

Fig. 12-1 —Four Cycle Internal Combustion Engine

that *no* air can be mixed in on the way to the cylinder. Again the mixture must be ignitable.

Refiners have found that there is a direct correlation between a gasoline's ability to meet those conditions and the RVP. Furthermore, they have found that the ideal RVP for gasoline varies with the seasons. In the dead of winter in a place like Bemidji, Minnesota, cold starts need a gasoline with a 13 psi RVP. During the dog days of August in Presidio, Texas, cars won't restart if the gasoline has a higher RVP than 8.5 psi.

Vapor Lock.   One other constraint on vapor pressure is worth mentioning—*vapor lock*. A combination of high altitudes and high temperatures can cause problems. At high altitudes, the atmospheric pressure is lower, and high RVP gasoline will tend to vaporize anywhere in the system. High temperatures aggravate the problem. The fuel pump is forced to pump a combination of vapor and liquid, when it is designed to handle only liquid. Consequently, the carburetor will be starved, and the engine will quit and won't start again until the temperature of the gasoline is lowered. That could take hours.

To avoid vapor lock, gasoline RVP is localized to accommodate the environmental conditions in the area, including seasonal temperature swings and barometric conditions.

Blending for Vapor Pressure.   So much for what cars need. How do refiners get there? If you look at the list of gasoline blending components in the table below, you'll see that all but two have RVP's below the limits mentioned above. The answer might jump right out at you: butanes are used as the *pressuring agent*.

Blending Components	RVP-psi
iC$_4$	71.0
nC$_4$	52.0
Reformate 94 RON	2.8
Reformate 100 RON	4.2
Light Hydrocrackate	3.9
Heavy Hydrocrackate	1.7
Alkylate	4.6
Straight–Run Gasoline	11.1
Straight–Run Naphtha	1.0
Cat Cracked Gasoline	4.4
Coker Gasoline	4.0

If you sat down to *design* the scheme for an industry to blend gasoline, you'd probably reject the thought that there would be enough butane around to be the marginal component used for all vapor pressure control. But amazingly, that's the way it turned out. Butanes are made in refineries as a by-product of conversion processes. They also are recovered from natural gas in gas processing plants. Somehow these two rather inelastic supplies end up providing all the butane that's needed for gasoline blending.

Getting down to nuts and bolts, the procedure for calculating the amount of butane needed to pressure up the gasoline involves only algebra and weighted averages. Vapor pressure calculations are not exactly related to the volumetric weighted averages, but for the purposes of this exposition, they're close enough. Suppose the RVP specification is 10 psi and you have a blend of five components. How much normal butane is needed?

Component	Barrels	RVP	Volume × RVP
Straight run gasoline	4,000	1.0	4,000
Reformate	6,000	2.8	16,800
Lt. Hydrocrackate	1,000	4.6	4,600
Cat cracked gasoline	8,000	4.4	35,200
	19,000		60,600
Normal butane	x	52	52x

For 10 psi RVP

$$10(19,000 + x) = 60,600 + 52x$$
$$190,000 + 10x = 60,600 + 52x$$
$$-52x + 10x = -129,400$$
$$x = 3,081 \text{ barrels of normal butane required}$$

Total gasoline production is $19,000 + 3,081 = 22,081$ bbl.

So the calculation is pretty simple, but some of the implications might be mentioned. Since the specification for RVP is higher in the winter than in the summer, the capacity to produce gasoline is higher in the winter. The higher the RVP spec, the more butane can be blended in and the more total volume made. Unfortunately, in most markets, except those like Miami Beach, the demand for gasoline is lower in the winter than the summer. The additional gasoline capacity does, however, give some flexibility to maximize distillate production.

Normal versus Isobutane. Why use normal butane instead of iso-butane to pressure gasoline? There are several good reasons. First of all, the RVP of normal is 19 psi less, which means that *more* normal can be blended in. Usually the price of the butanes is such that there is an incentive to blend as much butane as allowed. Secondly, there is another home for isobutane, alkylation. Very often, there isn't enough iso around to satisfy the alkylation needs, and some normal butane has to be processed in a *butane isomerization plant* (see Chapter XVI) to make the isobutane. Third, the market price of normal is often a little less than isobutane.

As an interesting footnote, do you recall ever watching a car gas tank being filled? Generally a wavy-looking vapor appears around the gas cap. That's butane escaping from the gasoline blend. If you recall correctly, you probably saw more vapors in the winter than in the summer. That's the result of the higher vapor pressure spec in the winter versus the summer.

Octane Number

Everybody who buys gasoline knows that high octane gasoline is better and more expensive. Some people know why it's better; hardly anyone knows why it's more expensive. This section is an attempt to demystify the subject.

Octane numbers are measures of whether a gasoline will *knock* in an engine. That's a fine definition, but it requires an explanation of another universally obscure phenomenon, knocking.

Knocking. It'll be helpful to refer back to Fig. 12-1, the diagram of the engine cycle. After the gasoline/air vapor is injected into the cylinder, the piston moves up to compress it. As the vapor is compressed, it heats up. (Ever feel the bottom of a bicycle pump after you've pumped up a tire. It's hot. Same effect as in an engine cylinder.) If the gasoline/air vapor is compressed enough, it will get hot enough to *self-ignite*, without the aid of a spark plug. If this happens before the piston reaches the top of the stroke, the engine will knock, that is, push against the crankshaft instead of with it. Knocking is usually perceived as a thud, ping, or knock coming from the engine.

Obviously knocking is something to be avoided since it not only works against the engine's motive power, but it's also tough on the mechanical parts. At the very early stages of engine development, it was discovered that the various types of gasoline components behaved differently. The key characteristic was the compression ratio. The *compression ratio* in Fig. 12-2 is simply the ratio of the volume at the bottom of the stroke to the volume at the top of the stroke. When measuring the octane number of a gasoline or gasoline component, a specific compression ratio is important:

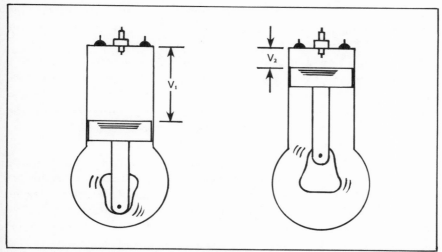

Fig. 12-2 —Compression Ratio Equals V_1: V_2

the one at which self-ignition will take place right at the top of the stroke. A series of guide numbers was devised to measure the compression ratio at which any gasoline component knocked. Iso-octane, C_8H_{18}, was defined as 100 octane gasoline. Normal heptane, C_7H_{16}, which knocks at a much lower compression ratio, was defined as zero octane gasoline. By using a test engine, any gasoline component can be matched with blends of iso-octane and normal heptane. The octane number equals the percent of iso-octane in the blend that knocks at the same compression ratio.

Testing for Knock. An explanation of the test procedure might help. A test engine is used that has a top which can be screwed up or down to vary the compression ratio. The gasoline whose octane number is to be measured is fed into the engine while the head is being turned down. At some point, knocking will occur, as detected either by ear or by the use of a detonation meter. After noting the compression ratio, the cylinder head is backed off. Two blends of iso-octane and normal heptane are prepared. From familiarity with the apparatus, blends that will knock with a higher and a lower compression ratio than that just measured can be concocted. The octane numbers of these blends are known by definition (the percent iso-octane). Each of these is run through the same test procedures with the critical compression ratio noted. By plotting the three data points, the octane number of the gasoline component can be read off the graph in Fig. 12-3.

For example, on a test engine a gasoline component knocks at a compression ratio of 8:1. Two test blends are made up, one with 88% isooctane (88 ON), the other with 96% iso-octane (96 ON). In the test engine, they knock at 7.2:1 and 8.4:1, respectively. From the chart, the octane number of the gasoline component must be 93.3 ON.

Octane Requirements.   So now you know what octane numbers measure. Why are they important? The design of an engine demands that the fuel behave in a certain way. The compression ratio of an engine determines the amount of power it can deliver. The higher the compression ratio, the longer the power stroke, the more powerful the engine. So different size cars have different engine designs, and therefore requirements for gasolines of different octane numbers. Put more simply, you don't get to vary the compression ratio of your car by turning the head up or down. So

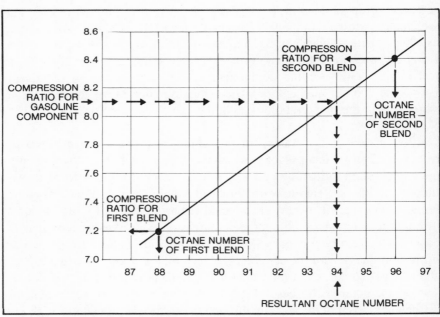

Fig. 12-3 —Plotting the Octane Number Test

you have to buy the quality of gasoline that accommodates what you have.

Types of Octane Numbers.   You need to know two more sets of nomenclature about octane numbers: the different kinds and their uses. First of all, the tests for octane numbers are run under two sets of conditions. The *Research Octane Number* (RON) test simulates driving under mild conditions; the *Motor Octane Number* (MON) test is run under more severe conditions and simulates operations under load or at high speeds. The two measures, RON and MON, give an indication of performance under the full range of conditions.

In the late 1960's there was a controversy between the U.S. Federal Trade Commission (FTC) and the U.S. refiners about posting octane numbers on gasoline pumps. The FTC wanted RON's posted. The refiners objected that RON didn't tell the whole story. The FTC considered posting RON and MON. "Too confusing," said the refiners. The FTC finally arrived at a compromise by ordering the following be posted on gasoline pumps:

$$\frac{\text{RON} + \text{MON}}{2}$$

That measure doesn't have any particular meaning other than the fact that the controversy was over.

The second piece of information about octane numbers is how they behave. When two gasoline components are mixed together, the RON's and MON's do not blend linearly. That is, the resulting RON and MON is not the volume averaged octane. However, and most fortunately, there is such a thing as a *blending octane number* for the RON and MON of every component which does blend linearly. The blending octane number is related to the true (test engine) number in a constant way, and is developed by experience. When references are made to RON's or MON's of components, they can mean either the true or blending octane number. Hereinafter, all references to octane numbers will mean the blending octane number, not the true number.

Blending for Octane Number.   An example may help pull these ideas together. Take the blend of gasoline in the previous example on the use of butane to achieve vapor pressure. Calculate the RON and MON of the blend.

	Barrels	MON	RON
Straight run gasoline	4,000	61.6	66.4
Reformate	6,000	84.4	94.0
Lt. Hydrocrackate	1,000	73.7	75.5
Cat cracked gasoline	8,000	76.8	92.3
Normal butane	3,081	92.0	93.0
	22,081		

The weighted average octane numbers for the 22,081 barrels are 78.1 MON and 87.4 RON.

Now calculate how much alkylate would have to be added to meet a minimum specification of 80.0 MON and 89.0 RON. Alkylate has octane numbers of 95.9 MON and 97.3 RON.

	Volume	MON	RON
Gasoline blend	22,081	78.1	87.4
Alkylate	Y	95.9	97.3
Specification (minimum)		80.0	89.0

In order for the 22,081 bbl of blended gasoline to meet the MON spec, the following alkylate must be added:

$$(22,081) (78.1) + Y (95.9) = (22,081 + Y) (80.0)$$
$$Y = 2,638 \text{ bbl}$$

The same kind of calculation is done for the alkylate to meet the RON spec:

$$(22,081) (87.4) + Y (97.3) = (22,081 + Y) (89.0)$$
$$Y = 4,257 \text{ bbl}$$

More barrels are needed to meet the RON spec, so that decides the volume, because both the MON and the RON specs are minimums. The extra MON is a giveaway.

A subtle problem has been encountered in this example. If 12,241 bbl of alkylate are added to meet the minimum octane specifications, the RVP specification is no longer met. By the use of two equations and two unknowns, the volume of butane and the volume of alkylate, the correct blend can be figured out. But it's more algebra and arithmetic than your attention span can stand here.

Leaded Gasoline

Lead was added to gasoline to greatly simplify blending for octane number. Lead, in the form of tetraethyl lead (TEL) or tetramethyl lead (TML), increases the octane number of gasoline without affecting any other properties, including vapor pressure.

TEL is a very toxic chemical, and in low concentration in the vapor form can induce violent illness or death. Because of this hazard, as late as the 1960's the Surgeon General of the United States (then part of the Executive Branch) set the maximum amount of TEL allowed in gasoline at 4.0 ml per gal. The Environmental Protection Agency succeeded to the authoritative position of the Surgeon General. Because of more intensive concern about air quality, in 1974 the EPA mandated a gradual phasedown of the lead content in gasoline, starting in 1975. As long as lead in the form of TEL or TML is still permitted in at least one grade of gasoline produced by a refiner, it will still be an important economic tool, albeit somewhat blunted.

It sounds contradictory, but lead is added to gasoline to *suppress* ignition. Remember that the lower the octane number, the more likely a gasoline is to knock, or self-ignite. Lead helps prevent the self-ignition.

A complicating factor in the use of lead is that the higher the concentration of lead additive, the less effective the last increment. In other words, the relationship between octane number and lead content is non-linear—it bends over.

In the table below and Fig. 12-4, you can see that the effect of lead is

Blending Component	MON			RON		
	Clear	*1.59g*	*3.17g*	*Clear*	*1.59g*	*3.17g*
iC₄	92.0	99.3	102.0	93.0	100.4	103.2
nC₄	92.0	98.8	101.5	93.0	99.9	102.5
Reformate 94 RON	84.4	89.3	91.2	94.0	99.4	100.8
Reformate 100 RON	88.2	92.3	93.7	100.0	103.1	104.0
Light Hydrocrackate	73.7	86.3	91.4	75.5	88.4	93.4
Heavy Hydrocrackate	75.6	84.6	87.9	79.0	88.3	92.2
Alkylate	95.9	101.9	103.4	97.3	102.0	104.0
Straight-Run Gasoline	61.6	73.7	80.6	66.4	77.3	83.5
Straight-Run Naphtha	58.7	72.5	78.2	62.3	73.5	79.3
Cat Cracked Gasoline	76.8	78.8	79.4	92.3	94.8	95.8
Coker Gasoline	76.6	80.6	82.1	85.5	90.7	93.0

unique for each type of gasoline component. Some components are more susceptible to *octane enhancement* than others. Furthermore, when the components are mixed together, the blend has its own octane enhancement curve that can be approximated from the curves of the components.

Usually the calculation technique calls for taking three points on each curve, the octane number with no TEL (called clear or neat), at 1.59 g per gal and 3.17 g per gal. (The latter two strange numbers relate to the weight basis equivalent to 1.5 and 3.0 ml or cc's of TEL). The weighted average octane number of the blend at each lead concentration is calculated. The blend's enhancement curve is approximated from the three points, and the amount of lead needed to achieve the octane specification can be read off.

To simplify the last step, the TEL and TML manufacturers have devised

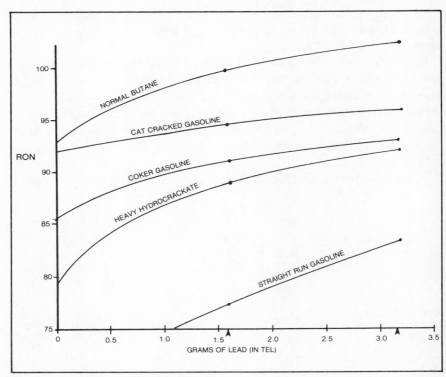

Fig. 12-4 —Effect of TEL in Gasoline

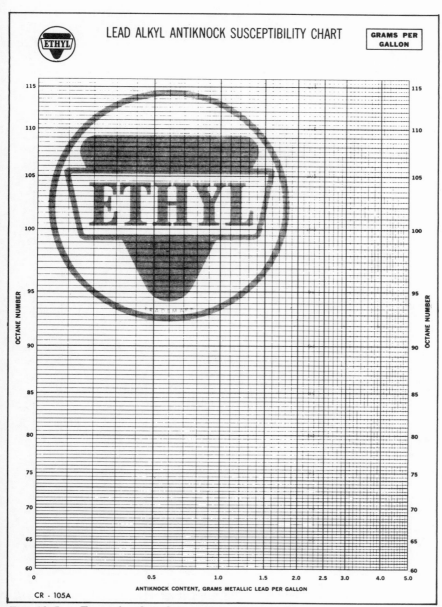

Fig. 12-5 —Example of an Octane Enhancement Chart

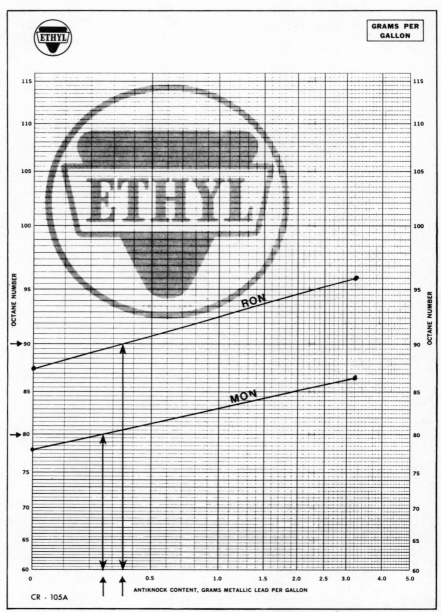

Fig. 12-6 —Plotting Research Octane Number and Motor Octane Number

a graph which can be used to read off octane numbers, given any two points, usually at 0 and 3.17 g. For simplicity, call it an octane enhancement chart as shown in Fig. 12-5.

As a demonstration, look again at the gasoline blending problem discussed above where alkylate was added to achieve the octane specification. Suppose the alternative was to add TEL instead of alkylate.

	Barrels	MON–0.0	MON–3.17	RON–0.0	RON–3.17
Straight run gasoline	4,000	61.6	80.6	66.4	83.5
Reformate	6,000	84.4	91.2	94.0	100.8
Lt. Hydrocrackate	1,000	73.7	91.4	75.5	93.4
Cat cracked gasoline	8,000	76.8	79.4	92.3	95.8
Normal butane	3,081	92.0	101.5	93.0	102.5
	22,081	78.1	86.5	87.4	95.8

Each pair of octane numbers for the blend is plotted on the octane enhancement chart (Fig. 12-6). The MON-O and MON-3.17 are connected by a straight line, as are the RON-O and RON-3.17. From those two lines, you can see that to meet the MON spec of 80, 0.25 g lead need be added; to meet the RON spec of 90, the lead requirement is 0.35 g. Therefore, the latter volume is used, and there is "give-away" on the MON.

Alcohols and Oxygenates

In the late 1970's when the EPA required the lead content to be reduced, refiners looked for other octane enhancers. From the petrochemicals industry came several alternatives: methanol, ethanol, TBA, and MTBE.

Methanol.   One of the oldest industrial chemicals around is methanol, CH_4OH. It is commonly known as wood alcohol because of the historic practice of making it by chemically treating fresh-cut lumber from hardwood trees. Since 1923, a more efficient process has been in use that starts with methane or naphtha. The source of the methane is usually natural gas, which is predominately methane.

An intermediate step in the process is the creation of synthesis gas, a mixture of carbon monoxide (CO) and hydrogen (H_2):

$$CH_4 + H_2O \rightarrow CO + 3H_2$$
(synthesis gas)

$$CO + 2H_2 \rightarrow CH_3OH$$
(methanol)

The formulas are deceptively simple. The process and hardware are complicated and expensive, and they involve expensive catalysts, vessels, and piping, with temperatures of 500–800°F and pressures of 4000–5000 psi.

Ethanol.   The alcohol most intimately familiar to everyone is ethanol, or ethyl alcohol, the primary constituent of whiskey. Like methanol, ethanol had natural beginnings—the fermentation of sugar in grapes (wine), potatoes (vodka), and grain and corn (whiskey). The present commercial synthetic process for making ethanol was developed in 1919. It involves the direct hydration (addition of water) of ethylene:

$$C_2H_4 + H_2O \rightarrow CH_3CH_2OH$$
(ethylene)   (water)   (ethanol)

In the U.S., synthetic ethanol cannot be sold for human consumption, excluding it from the alcoholic beveragel industry.

TBA.   Tertiary butyl alcohol is available from several processes. One of the processes for making propylene oxide gives TBA as a by-product. It can also be made by reacting normal or isobutylene with water or by starting with propylene and isobutane. All of them yield an alcohol, $(CH_3)_3COH$.

The use of methanol as a gasoline blending component increased refiners' interest in TBA. If a gasoline blend includes methanol, contact with small amounts of water can cause phase separation, i.e., the components "unblend." When heavier alcohols, such as TBA, are mixed with methanol, the *water tolerance*—the amount of water that can be tolerated before phase separation—is increased. TBA acts as a *cosolvent;* it helps methanol stay dissolved whenever water is present.

MTBE.   Methyl tertiary butyl ether (with a name like that, no wonder it's called MTBE) is an *oxygenate*. That is in contrast to methanol, ethanol, and TBA, which are alcohols having the characteristic -OH at the end of their formulas. But MTBE does have an oxygen atom in its formula, $(CH_3)_3COCH_3$.

MTBE is produced by reacting isobutylene with methanol over a catalyst. The feed to an MTBE plant is *mixed C_4's* (butanes, normal butylenes, and isobutylene). Almost all the isobutylene is reacted out in the process. In some situations, this coincident phenomenon can be an added benefit (the need for an iso-free stream for sale, for example.)

Handling Characteristics.   The problems of water contamination have already been alluded to. They can be severe. To illustrate, think about two glasses—one with an ounce and a half of scotch (primarily ethanol and water) and the other with the same amount of gasoline. Pour about four ounces of water into each and stir. The oil and water separate, of course, as soon as the stirring stops. The scotch and water mix nicely.

Now mix the contents of the two glasses into a third and shake well. (What a waste, just for science.) The liquids will separate again but the scotch will stay in the water, not mix in the gasoline. In the U.S., the major portion of the volume of gasoline is moved in pipelines. Those systems have never been fitted to completely exclude all water. Consequently, gasoline blended with alcohols can *phase separate* in pipelines. Therefore, it must be handled by some other means of transort. MTBE, because of its different chemical structure, has very low affinity for water and is not as limited in use as the alcohols.

Blending Characteristics.   The interest in the alcohols, after all, is in the relationship between their cost and their gasoline blending characteristics—more specifically their octane enhancement. Alas, the story is not as straightforward as, say, alkylate or reformate. The alcohols and oxygenates perform in a discontinuous way as they are added to a gasoline blend. For example, small additions (up to 2% or 3%) of the alcohols raise the RVP of the blend sharply. Beyond that (up to 5, 10, or 15% volume) there is no change at all. Over the range of these higher concentrations, methanol increases the RVP of the blend by about a constant 3 RVP, ethanol about 0.7 RVP, and TBA about 2 RVP.

Response to the addition of lead to the blend is erratic as well and is dependent on the other components of the gasoline blend. In many cases, there is a negative response to the addition of lead. Inasmuch as the alcohols and oxygenates are being developed for unleaded gasoline blending, the erratic lead effects are not of much importance.

Gasoline Blending and Its
Impact on Operations

There is no pretending that optimizing the blending of gasoline is simple. It is very complex, especially now that several grades of gasoline cannot be leaded. Consider for a moment the ever-increasing levels of complexity.

a. Given the demands for three grades of gasoline and the availability of components, blend up the requirements—with no leftovers.

b. Now consider varying the operating conditions of some of the process units. Change the severity on the reformers to adjust yields versus octane numbers; increase the temperature in the cat cracker to generate more olefins and ultimately more alkylate, etc.

c. Finally, consider diverting streams in and out of units. Send cat cracked light gas oil to be blended to furnace oil rather than hydrocracked; blend butylenes directly into motor gasoline instead of alkylating; cut the bottoms off the straight-run naphtha (reformer feed), making more kerosene/turbine fuel.

The most successful technique for coping with all these variables is the use of linear programming on a large computer to simulate refinery operations. The intakes, outturns, capacities, and costs of each operation, from distilling to blending, are described using several equations and numerical values. The crude oil availabilities and costs, and the product demands and prices are detailed. The linear programming technique will find the solution to the equations (there are usually many) that makes the most profit.

The computer is necessary because of the thousands of calculations that are necessary to find the optimal solution. But even this is only an approximation, for several reasons:

a. The data fed into the models are estimates of the process unit yields. Depending on any number of things (time since the last shutdown, catalyst activity, air temperature, cooling water temperature, etc.) the yields could vary.

b. The crude composition could vary.

c. The demands and prices could vary.

Further, the inevitable unscheduled shutdowns in some part of the refinery will interrupt orderly flow. Nonetheless, as an analytical technique to develop a model or a plan, the linear program is an invaluable tool.

Conclusion. The subject of gasoline blending brings into focus most of the upstream operations in a refinery. The rudiments of octane satisfaction are simple enough, but optimizing gasoline blending implies optimizing the entire refinery.

EXERCISES

1. Define the following terms:

vapor pressure	knocking
RVP	compression ratio
power stroke	RON and MON
vapor lock	leaded gasoline
pressuring agent	octane enhancement

2. Calculate the amount of normal butane needed to produce a 12.5 psi RVP for a mixture of 2,730 bbl of straight run gasoline; 2,490 barrels of 94 RON reformate; 6,100 bbl of heavy hydrocrackate; and 3,600 bbl of cat cracked gasoline. How much TEL must be added to produce a 97.0 RON gasoline?

What problem occurs with that much TEL? What three things happen to the gasoline pool if the heavy hydrocrackate is run through the reformer before being used as a gasoline component? Assume a yield of 85% reformate and octane numbers and vapor pressures from the tables in this chapter.

XIII.

DISTILLATE FUELS

Warmth, warmth, more warmth! for we are dying of cold and not darkness. It is not the night that kills, but the frost.
—*The Tragic Sense of Life*, Miguel de Unamuno

Diesel fuel and furnace oil are the two most widely used fuels made out of the light gas oil range streams in a refinery. You might even say they are the *one* most widely used fuel because in most places diesel fuel and furnace oil are the same thing and are sold out of the same tank.

Diesel Engines

A cut-away view of a diesel engine looks a lot like a gasoline engine. The one noticeable difference is that there is no spark plug. The mechanics of a diesel engine depend on the fuel igniting itself—the very thing that was important to avoid in a gasoline engine.

The control of self-ignition depends very acutely on timing. Unlike gasoline engines, air is not mixed together with the fuel before it is injected into the cylinder. Only air is injected. As the piston moves to the top of the compression stroke, the air gets hotter and hotter as it is compressed. Just as the piston reaches the top of the stroke, the diesel fuel is injected into the cylinder. On contact with the superheated air, it ignites and causes the piston to begin the power stroke.

Several distinct phases happen as the fuel is injected into the cylinder. First, the fuel is injected in the liquid form, although it may be sprayed to disperse it. As the liquid hits the superheated air, it vaporizes and is raised to the self-ignition temperature. The second phase begins as combustion occurs and starts igniting the surrounding liquid and vapor. The power stroke begins. Finally as the rest of the liquid is pumped into the cylinder, it is also ignited, maintaining or increasing the pressure on the piston. All this happens in about a thousandth of a second.

The precision required of the timing mechanism becomes apparent. If the fuel is not injected at the correct moment and rate, too much fuel could be ignited at once, causing an explosive power stroke rather than a controlled surge. Fuel injection apparatus for diesels are much more precisely engineered than those for gasoline engines. Because the diesel fuel is injected at the top of the stroke, operating pressures of the fuel injection system can be in the range of 2000–10,000 psi.

Diesel Fuel

Many of the properties that made gasoline better make diesel fuel worse. Remember that self-ignition of gasoline was a phenomenon to be avoided. For diesel fuel it is essential: the most important characteristic is the ignition quality. A measure of this is the *cetane number*. Reminiscent of octane number and its derivation, cetane number of a diesel fuel corresponds to the percent of cetane ($C_{16}H_{34}$) in a mixture of cetane and alpha-methylnaphthalene. When that mixture has the same ignition characteristics in a test engine as the diesel fuel, the diesel fuel has a cetane number equal to that percent cetane.

The cetane number is partially a function of the PONA (paraffin, olefin, naphthene, aromatic) content. In gasoline engines, aromatic compounds resist self-ignition; paraffins have low self-ignition temperatures. In diesel engines, low self-ignition makes paraffin fuels more desirable, aromatics (prevalent in cracked gas oils) less attractive.

Like gasoline, there are several grades of diesel fuel. Regular diesel fuel runs about 40-45 cetane; premium diesel runs 45-50 cetane. The latter fuel usually has more *lighter* range, more volatile fractions in it, and therefore is better for cold starts. The lighter fractions tend to have lower carbon-hydrogen ratios, which lead to less formation of smoke under severe conditions. As a consequence, premium diesel is often a preferred fuel for city buses.

Blending Components.   All the light gas oils are candidates for diesel fuel blending, but some are better than others. Straight run light gas oil is generally very paraffinic, and usually has a cetane number in the 50–55 range. The cracked light gas oils generally have high olefinic, aromatic, and naphthenic concentrations and therefore are usually in the 32-35 range. Kerosene is usually about 55 cetane. From this wide range of components, blends meeting the cetane specs are relatively easy to achieve.

Furnace Oil

The gas oil range hydrocarbons have a number of physical characteristics that have made them the most popular petroleum heating oil. First of all they carry more heating capability (*calorific content*) than the lighter hydrocarbons such as kerosene, naphtha, or LPG. Second, they are cheaper to transport than natural gas or LPG because pressure equipment is not needed. Third, they are not as susceptible to accidental or explosive ignition as naphtha. Fourth, they are easier to burn than residual fuels because they do not have to be heated before they are injected into the fire box. Finally, they are easier to render pollution-free than residual fuels because of their less complex chemical constituency.

For all these reasons, furnace oil has been used extensively for residential and commercial heating and, to a lesser extent, for light and heavy industrial furnace needs. Furnace oil has several synonyms: *number 2 fuel, distillate fuel, Two Oil.*

Specifications for furnace oil cater primarily to the limitations in the design of residential heating systems. From a safety point of view, *flash point* is important; because furnace oil is transported and stored in unheated equipment, *pour point* is important.

Flash Point. The lowest temperature at which enough vapors are given off to form a combustible mixture with air is called the flash point. The flash point is therefore a measure of both volatility and inflammability. For furnace oil it is set at a limit that protects against vapors leaving the tank and finding a source of ignition other than in the furnace. (Many liability suits have resulted from gasoline accidentally being mixed with furnace oil. The vapors leak out of basement storage tanks, find the pilot light on a nearby hot water heater, and explode.)

Pour Point. The ability of a petroleum product to flow at low temperatures is measured by its pour point. When petroleum products are cooled, a point is reached when some of the constituents begin to solidify. If cooling continues, eventually the oil will not flow. The pour point is 5°F higher than the temperature at which the product stops flowing. The pour point of furnace oil is about −10°F.

Blending Furnace Oil. It's no coincidence, but it is certainly fortunate that the products that are suited to be fuels for the internal combustion diesel fuel also fit the requirements of residential heating systems. There is a lot of latitude in putting the various cracked light gas oil streams into furnace oil, more so than into diesel fuel, because of the cetane number. But within that constraint, most refiners blend, ship, store, and sell one petroleum product which satisfies the needs of two very different markets: heating and internal combustion.

Given the cetane numbers of straight run light gas oil, cracked light gas oil, and kerosene, plus whatever other miscellaneous distillate range streams are around, it is a simple matter to calculate the blend that meets the minimum cetane number. To achieve the pour point, kerosene is usually added, but often at an economic penalty (jet fuel sells for more than distillate). Too much kerosene could cause the minimum flash point spec to be passed. The use of anything lighter than kerosene will cause flash point problems for sure. Specialty chemical companies are all too happy to sell additives that control pour points.

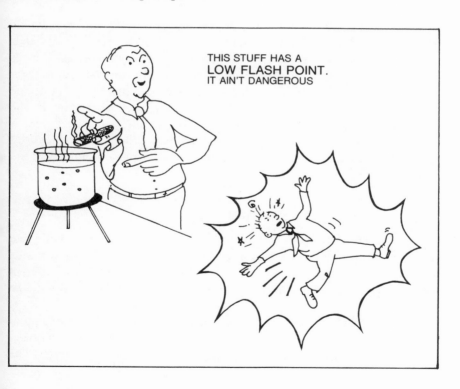

XIV.

ASPHALT AND RESIDUAL FUEL

Muddy, ill-seeming, thick, bereft of beauty.
—*The Taming of the Shrew*, Shakespeare

A refiner has two ready alternatives to handle the bottom of the barrel: residual fuel and asphalt. His choice will depend on the characteristics of the crude oil he runs and the access to markets.

ASPHALT

At the heavy end of crudes are varying amounts of *asphaltenes*, which are very complex molecules. Asphaltenes are polyaromatic compounds with high carbon to hydrogen ratios because of the way the aromatic rings are so densely connected. An organic chemist might smugly refer to them as 2,3 trimethyl chicken wire. The number of carbons in the molecule usually exceeds 50. When the asphaltene content is high enough, the material makes a good, strong, stable binder: asphalt.

Asphalts can be categorized into four groups, although the categories overlap. There are straight run asphalts, blown asphalts, cutbacks, and emulsions.

Straight Run Asphalts

Deep flashing of the straight run residue of an asphaltic crude will yield flasher bottoms that can be used directly as an asphalt. The bottoms are a black or dark brown material, very viscous (thick) or solid at ambient temperatures. The higher the flasher temperature, the more viscous the asphalt.

There are two usual tests for grading this type of asphalt: softening point and penetration. The *softening point* is the temperature at which an object with a standardized weight and shape will start to sink into the asphalt. The

108

most popular test method uses a steel ball. Commercial grades of asphalt have softening points somewhere around 80-340°F.

The hardness of an asphalt, once it's applied, is measured by its *penetration*. The test apparatus for penetration has a long needle with a weight on top. The depth to which the needle penetrates the asphalt over a standard period of time at a given temperature is the measure of penetration. Very hard asphalts are zero penetration, while the softer ones range up to 250.

A wide variety of grades of asphalt can be dispensed at a refinery by making *blocked-out runs;* for a period of time, the flasher can be run to fill a tank with 40-50 penetration asphalt. Operating conditions on the flasher

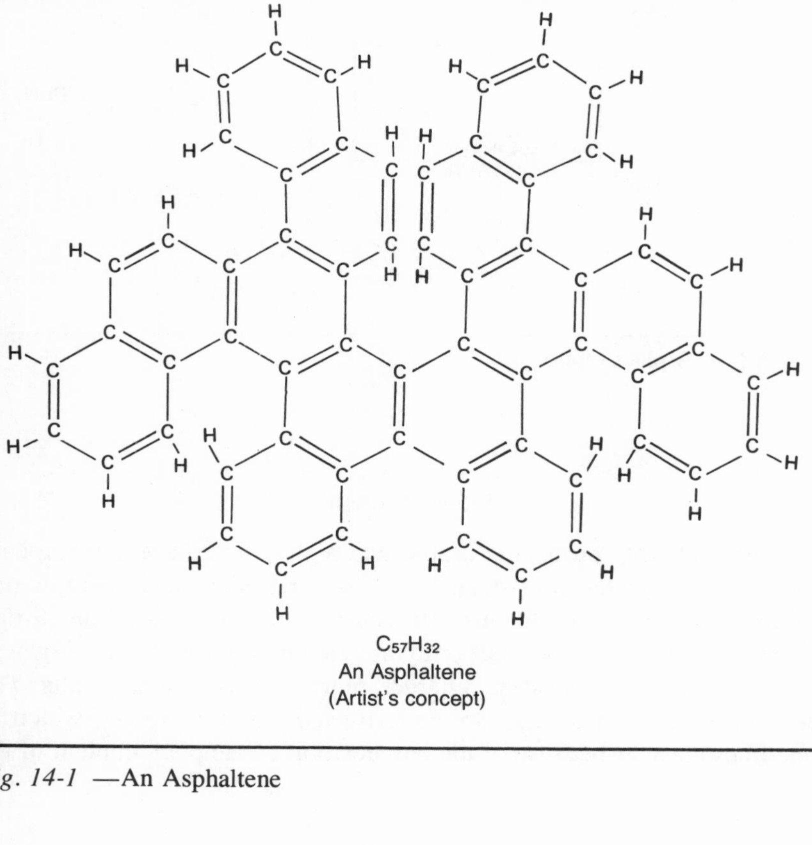

$C_{57}H_{32}$
An Asphaltene
(Artist's concept)

Fig. 14-1 —An Asphaltene

can then be changed to make a 250 penetration grade, running it into another tank. By blending these two grades directly into trucks, tank cars, ships, or barges, any grade in between 40 and 250 penetration can be produced.

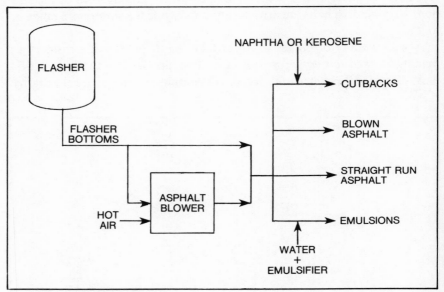

Fig. 14-2 —Asphalt Options

Blown Asphalt

An alternative to producing asphalt via high temperature flashing is to chemically change the consistency of the softer grades in a blower. In this unit hot air is blown into the asphalt, causing a chemical reaction. Either the oxygen bonds to the asphalt or hydrogen combines with the oxygen to make water which evaporates. A harder, more rubbery asphalt results. The lower penetration asphalts can be achieved via this route, which is sometimes cheaper because of the composition of the pitch content of the crude.

Cutbacks

The application of asphalt requires that the materials flow. In road construction, for example, the asphalt must completely surround the aggregate (gravel) and sand with which it is mixed. Normally asphalt is mixed and applied at heated temperatures. To reduce the severe equipment requirements due to temperature at the construction sites and at the same time achieve a hard durable product, sometimes a thinner is added to the asphalt. The thinner, or diluent, will soften the asphalt and permit lower application temperatures. After the asphalt has been applied, the diluent will evaporate, leaving the hard, durable asphalt behind.

The rate at which the diluent evaporates, usually in terms of days, will depend on the diluent used to cut the asphalt back. An asphalt that needs to *cure* (achieve its final, stable state) rapidly can be cut with a very light diluent, like naphtha. If a longer curing time is acceptable, kerosene can be used. The terms RC Cutback and MC Cutback (Rapid Curing and Medium Curing Cutbacks) generally denote the type of diluent used.

The use of cutbacks has always been somewhat problematic. The diluent just passes through the system and doesn't end up in the finished product. That's expensive. Secondly, the evaporation of the diluent is a source of air pollution.

Emulsions

To achieve the same flexibility in field equipment but avoid the cutback problems, emulsions have been developed. Emulsions are a mixture of about 50–70% asphalt and 30–50% water. To keep the asphalt and water mixed, an emulsifying agent which is something like soap is added. The agent, though much more expensive per gallon than the asphalt, comprises only about one percent.

After application, the water evaporates just like the cutback diluents, leaving the hard asphalt behind. Emulsions are generally applied at lower temperatures than any of the grades mentioned before, since the addition of water to facilitate handling is pretty cheap.

Asphalts for road construction are usually the softer grades, 200–300. Industrial asphalts generally use harder grades, 40–50 penetration. Often clay is mixed in with the asphalts for roofing (shingles, tar paper, and felt) and flooring materials (damp-proofing coatings, tiles).

Some clever, inquisitive engineer discovered asphalt particles in emulsions are by nature negatively charged; they are called *anionic* emulsions. Many aggregates that are used as mixes with asphalt are also by nature negatively charged. Sometimes it is difficult to completely coat the aggregate with the asphalt emulsion, especially when they are mixed in damp weather when electricity can really flow. To combat this problem, special emulsions have been developed in which the asphalt particles are positively charged. These emulsions, called *cationics,* readily grab the negatively charged aggregates. Asphalt emulsions used for road construction are now mostly cationics.

RESIDUAL FUELS

Residual fuel is what's left over after you make everything you want. That's why it's called *residual*. In many non-U.S. refineries, residual fuel is no more than the crude distilling unit bottoms, called *long residue*. Most U.S. refineries use long residue as feed to a flasher. In that case, flasher bottoms (pitch) are the primary ingredient of *resid*.

Residual has always commanded lower prices than the other fuels for two reasons, one physical and the other commercial. Moving and consuming resid requires special equipment. The pipelines, transportation, and storage must be heated so that the residual fuel doesn't solidify. Secondly but more importantly, almost all resid is a leftover from producing the other products. Its supply elasticity is nearly zero.

Blending Resid

Because resid is a by-product, the specs for marketable product are very loose. The most important spec is viscosity, a measure of how much a fluid resists flowing as shown in Fig. 14-3. (Maple syrup is more viscous than water.) The standard unit of measure for viscosity is *centistokes*. Since viscosity changes with temperature (hot syrup flows more easily than cold), the centistokes are measured at (usually) either 80°C or 100°C. Flasher bottoms generally need to have some kind of a diluent or *cutter stock* added to meet the maximum viscosity spec. Typically, the diluent is a heavy stream of low viscosity and a relatively low value as feed to a conversion unit (cracker). Cat cracked gas oil is usually used, if available.

Another specification is sulfur content. The allowable sulfur emissions from industrial fueling vary widely by geographic area and by the requirements of the consumer. Residual fuels are either high sulfur or low sulfur, with the break somewhere around 0.5 to 1.0% by weight sulfur. The actual content of the resid is primarily a consequence of the nature of the crude oil being run in the refinery. The technology for removing sulfur from residual fuel requires the expensive machinery described in the next chapter. As a consequence, the price differential between low sulfur and high sulfur reflects in some fashion the capital cost of desulfurization, as well as the increasing demand for lower sulfur residual fuels.

Finally, like distillate fuels, the flash point of residual fuels sometimes binds. Because residual fuels have to be heated to be pumped, the flash point is even more critical than for distillates. At the same time, residual fuels tend to be a sump where a lot of miscellaneous streams can be dumped. Flash point is usually the limiting factor for what can be hidden in residuals.

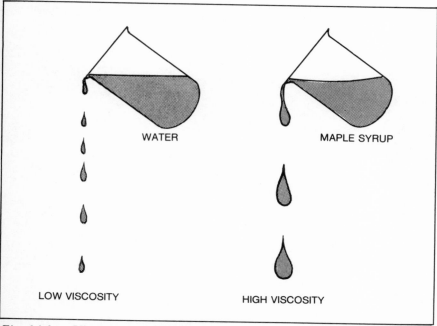

WATER

MAPLE SYRUP

LOW VISCOSITY

HIGH VISCOSITY

Fig. 14-3 —Viscosity is Resistance to Flow

A technique for increasing the sales of residual fuels made from high sulfur crudes became fashionable in the Caribbean refineries. The sulfur can be removed from the gas oil range streams by relatively inexpensive hydrotreating processes, as described in the next chapter. By blending the high sulfur flasher bottoms with the desulfurized gas oils, the resulting residual can meet a low sulfur spec. In some circles this activity was described as "solution to pollution is dilution."

XV.

HYDROGEN, HYDROTREATING AND SULFUR PLANTS

Neither to smell rank nor to smell sweet pleases me.
—*Epigrams,* Decimus Magnus Ausonius

All sorts of contaminants are found in crude oil. As the petroleum fractions travel through the refinery processing units, these impurities can have detrimental effects on the equipment, the catalysts, and the quality of the finished product. Furthermore, there may be legal or paralegal limits on the content of some impurities, like sulfur.

Hydrotreating does an effective job in removing many of the contaminants from many of the streams. Hydrogen is a vital input to the hydrotreating process.

Hydrotreating

Refinery streams that have C_6 and heavier hydrocarbons in them are likely to have some sulfur compounds as well. The sulfur can be attached or imbedded anywhere in the molecule and therefore is chemically a part of the stream. Hydrotreating has been used successfully to break the sulfur away.

In the hydrotreating process, the stream is mixed with hydrogen and heated to 500-800° F. The oil combined with the hydrogen is then charged to a vessel filled with a catalyst in pellet form. In the presence of the catalyst, several reactions take place:

a. The hydrogen combines with the sulfur atoms to form hydrogen sulfide (H_2S).

b. Some nitrogen compounds are converted to ammonia.

c. Any metals entrained in the oil are deposited on the catalyst.

d. Some of the olefins, aromatics, or naphthenes get hydrogen saturated and some cracking takes place, causing the creation of some methane, ethane, propane, and butanes.

The stream coming from the reactor is sent to a flash tank where most of

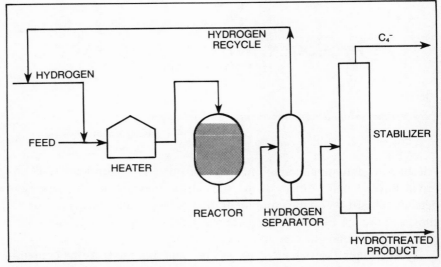

Fig. 15-1 —Hydrotreater

the propane and lighter, including the H_2S and the tiny bit of ammonia, go overhead. To completely strip out these light ends, a small fractionator is generally tacked onto the tail end of the process.

The importance of hydrotreating has been gradually increasing for many years for two reasons:

a. Removal of sulfur and metals is important protection for the catalysts in reformers, cat crackers, and hydrocrackers.

b. Air quality regulations are continually lowering the allowable sulfur content in fuel oils, calling for desulfurization of distillates and jet fuels.

Residual Hydrotreating.   Residual fuels are likewise coming under environmental pressure, and, somewhat belatedly, residual desulfurization facilities have been commercially developed. Although the flow diagrams are similar to lighter stream hydrotreaters, the hardware and yields are different. Residual streams have much lower hydrogen-to-carbon ratios. Despite the presence of excess hydrogen in the reactor, high pressures must be maintained to minimize coke formation. So a residual hydrotreater must be built as sturdily as a hydrocracker, an expensive proposition.

The yield from hydrotreating residuals has higher light end production. In those large molecules, especially the "trimethyl chickenwire" compounds, sulfur, nitrogen, and metals cannot be removed without literally destroying the molecule to spring the imbedded pollutant. In the process the smaller molecules result.

Jet Fuel Hydrotreating.   Hydrotreating can be used to improve the burning characteristics of distillates, especially jet fuel. The kerosene fraction can contain a large percentage of aromatic compounds which have higher carbon-to-hydrogen ratios. When these compounds burn, the deficiency of hydrogen can cause smoke. As a matter of fact, one of the specifications on jet fuel is the *smoke point*.

The apparatus used to measure smoke point is similar to a kerosene lantern. A reservoir of fuel is fitted with a wick that can be cranked up or down to vary the length and the flame size. The smoke point is a measure of how far the wick can be cranked up before smoke is visible above the flame and is equal to the wick length, in millimeters.

A kerosene with a low smoke point can be improved by hydrotreating. During the process, the aromatic rings get saturated with hydrogen, converting them to naphthenes which are cleaner burning compounds.

Pyrolysis Gas Hydrotreating.   One of the coproducts of making ethylene from naphtha or gas oil is pyrolysis gasoline (Chapter XIX). This stream is high in *diolefin* content, that is, the molecules have two sets of double bonded carbons. Often, pyrolysis gas is only suitable for gasoline blending in small concentrations. It smells bad, has a funny color, and forms gum in carburetors.

Hydrotreating saturates the double bonds, eliminating most of the undesirable characteristics. Some of the octane rating may decline during hydrotreating because of benzene ring saturation.

Hydrogen Plant

The normal source of hydrogen in a refinery is the cat reformer. The light ends of the reformer column contain a high ratio of hydrogen to methane, so the stream is de-ethanized or depropanized to get a high concentration of hydrogen stream.

Sometimes the reformer hydrogen cannot satisfy all the hydrogen requirements in a refinery. This might be true if there is a hydrocracker in operation. Hydrogen can be produced on purpose in a plant called a *Stream Methane Reformer* (SMR) as shown in Fig. 15-2.

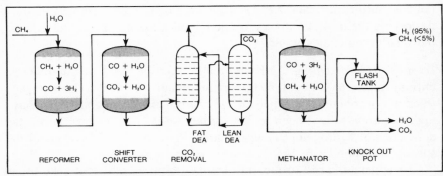

Fig. 15-2 —Steam Methane Reformer

When the design engineers were searching for ways to make hydrogen, chemical compounds with a high proportion of hydrogen were considered in order to waste as little as possible of the remaining material or the energy used to process it. The two compounds that were finally used were almost too obvious, methane (CH_4) and water (H_2O).

The trick in the SMR is to spring as much of the hydrogen from methane and water as possible, but with a minimum amount of energy (fuel) consumed in the process. With the aid of some very useful catalysts, the SMR operates in four stages:

1. *Reforming.* Methane and steam (the H_2O) are mixed and passed over a catalyst at 1500° F, resulting in the formation of carbon monoxide and hydrogen:

$$CH_4 + H_2O \rightarrow CO + 3H_2$$

2. *Shift Conversion.* Not content with the hydrogen already formed, the process exploits the carbon monoxide. More steam is added over another catalyst at 650° F to form carbon dioxide and hydrogen:

$$CO + H_2O \rightarrow CO_2 + H_2$$

3. *Gas Purification.* In order to provide a concentrated hydrogen stream, the carbon dioxide is separated from the hydrogen by a solvent extraction process.

4. *Methanation.* Since the presence of any carbon monoxide or dioxide in the hydrogen stream can mess up some of the applications, a cleanup step converts them to methane. A catalyst at 800° F is used:

$$CO + 3H_2 \rightarrow CH_4 + H_2O$$
$$CO_2 + 4H_2 \rightarrow CH_4 + 2H_2O$$

In some places, a sulfur free methane stream (natural gas) is not available. Heavier hydrocarbons such as propane or naphtha can be substituted. The equipment design and catalysts are different and the process is less fuel efficient, but it works.

Sulfur Facilities

Hydrotreating creates hydrogen sulfide (H_2S) streams—a deadly, toxic gas that needs disposal. The usual process involves two steps: first, the removal of the H_2S stream from the hydrocarbon streams; second, the conversion of the lethal H_2S to elemental sulfur, a harmless chemical.

H_2S Recovery.   Until about 1970, most of the H_2S in refineries was used, along with the other light ends, as refinery fuel. When it is burned in the furnaces, sulfur dioxide (SO_2) is formed. Air quality regulations now limit SO_2 emissions to the extent that most of the H_2S must be kept out of the fuel systems.

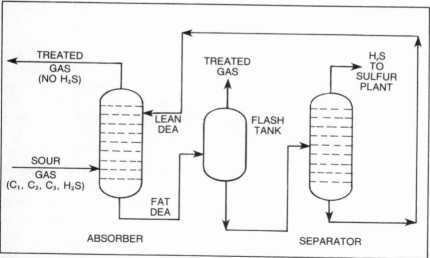

Fig. 15-3 —Diethanolamine Treater (H_2S Recovery)

Recovery of the H_2S can be done by a number of different chemical processes. The most widely used is solvent extraction using diethanolamine (DEA). A mixture of DEA and water is pumped down through a vessel with trays or packing in it. The gas stream containing the H_2S is injected from the bottom. As the streams circulate, the DEA will selectively absorb the H_2S gas. The rich DEA is then fractionated to separate the H_2S which is sent to a sulfur recovery plant. The stripped DEA is recycled. This process is reminiscent of the lean oil/fat oil demethanation process described much earlier in Chapter VII on gas plants except that the DEA selectively picks up the H_2S, but not any hydrocarbons.

Sulfur Recovery.   The conversion of H_2S to plain sulfur is done in a process first developed by a German named Claus in 1885. There are variations on the process today suited to various H_2S/hydrocarbon ratios, but they mostly use a basic two-step, split-stream process.

1. *Combustion.* Part of the H_2S stream is burned in a furnace, producing SO_2, water, and sulfur. The sulfur is formed because the air (oxygen) admitted to the furnace is limited to one third the amount needed to make all SO_2.

$$2H_2S + 2O_2 \rightarrow SO_2 + S + 2H_2O$$

2. *Reaction.* The remainder of the H_2S is mixed with the combustion products and passed over a catalyst. The H_2S reacts with the SO_2 to form sulfur.

$$2H_2S + SO_2 \rightarrow 3S + H_2O$$

The sulfur drops out of the reaction vessel in the molten (melted) form. Most refineries store and ship sulfur in the molten state, although some companies store sulfur by pouring it on the ground (in forms), letting it solidify. Sulfur can be stored indefinitely in this dry state, called a pile.

Claus plants convert about 90% to 93% of the H_2S to sulfur. Local environmental conditions may permit the balance of the H_2S, known as the tail gas, to be burned in the refinery fuel system. Alternatively, the tail gas could be processed for high percentage H_2S removal in more elaborate processes like Sulfreen, Stretford, or SCOT (Shell Claus Offgas Treating).

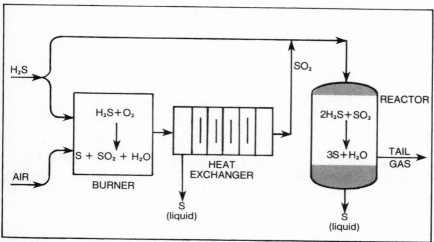

Fig. 15-4 —Claus Sulfur Plant

EXERCISES

1. Identify which of these streams are feeds, products, or internal streams in hydrotreating, DEA removal, Claus plants, and an SMR.

H_2S	CH_4
S	H_2
CO	O_2
CO_2	SO_2

XVI.

ISOMERIZATION

"My dear Bilbo! You are not the same hobbit that you were."
— *The Hobbitt*, J.R.R. Tolkien

Isom plants are molecule rearrangers somewhat like the cat reformer except that they only convert normal paraffins to isoparaffins. Curiously, the C4 isomerization unit is built for very different reasons than the C5/C6 isomerization unit.

BUTANE ISOMERIZATION

A refinery that has an alkylation plant is not likely to have exactly enough isobutane to match the propylene and butylene needs. If there is a hydrocracker in the refinery, there is likely to be surplus isobutane which is probably blended off to gasoline. If there is not, there is a need to supplement the supply. The choices are usually two—buy it or make it on a butane isomerization (BI) plant.

The Process

The feed to the BI plant is normal butane or mixed butanes (iso and normal), which are sometimes called field grade butanes if they come from a gas processing plant. The butanes should not have any trace of olefins which would deactivate the catalyst.

The butanes are fed to a feed preparation column where isobutane is removed; the high purity normal butane is then mixed with a small amount of hydrogen and chloride and charged to a reactor containing a platinum catalyst. The catalyst causes the normal butane to reform itself into its isomer, isobutane.

The stream coming from the reactor contains about 60% isobutane, 40% normal butane, and a minor amount of propane and lighter. In a

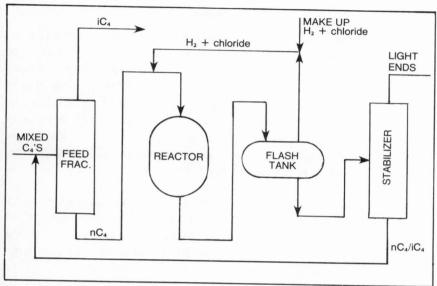

Fig. 16-1 —Butane Isomerization Plant

fractionator, the latter are split out and sent to the fuel system; the butanes are recycled to the feed fractionator so that the normal butane can be rerun.

Yields. When the yields are figured on a net basis, the isobutane outturn slightly exceeds the normal butane feed (the volume/weight trick again.) In essence, it's normal butane in, isobutane out, and that's it.

C_5/C_6 ISOMERIZATION

For a refinery that has problems meeting the octane number of gasoline and has a lot of straight run gasoline around, C_5/C_6 isomerization has appeal. Normal pentane, which has an RON of 62, can be converted to isopentane with an RON of 92. Hexanes go from an incredibly low 25 RON to about 75; a typical mixture of iso and normal pentanes and hexanes can be upgraded from 73 to 91 RON.

The Process

Like the BI plant, the C_5/C_6 Isom plant may have a feed fractionator that concentrates the normal pentanes and hexanes, rejecting the isomers. The normal paraffins are mixed with a small amount of hydrogen and organic chlorides and charged to a reactor. The catalyst will cause conversion of about half the feed to isomers (*isomerate*), so the reactor product can be fractionated to recycle the normal pentanes to extinction. Since the hexanes boil at a higher temperature than the normal pentane, but the isopentane boils at a lower temperature, the cost of the additional hexane splitter sometimes precludes recycling the normal hexane.

Yields.   The C_5/C_6 isomerization is a little more complex than C_4 isomerization. So about 2-3% light ends, C_4 and lighter, get created in the process. Depending on the amount of recycle, the octane number of the isomerate can be varied from 80 to 91 RON, with the cost of energy (fractionation, pumping) increasing with the octane number.

Review   Butane isomerization is used to satisfy the feed needs of alkylation by converting normal butane to isobutane. C_5/C_6 isomerization is a method of increasing the octane number of the light gasoline components normal pentane and normal hexane, which are found in abundance in light straight run gasoline.

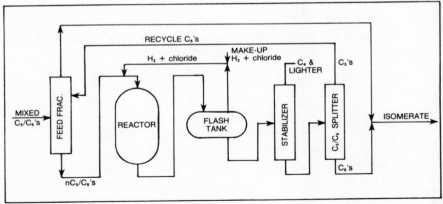

Fig. 16-2 —C_5/C_6 Isom Plant

XVII.

SOLVENT RECOVERY OF AROMATICS

We, only we, are left.

—"Rugby Chapel," Matthew Arnold

In 1907 a man named Edeleanu developed a process by which aromatic compounds could be preferentially removed from a hydrocarbon mixture. The process worked even in cases where the boiling point of the aromatic compounds were the same as the other compounds in the mixture. Distillation under those conditions wouldn't work, so Edeleanu's technique of using a solvent was a major breakthrough.

Applications

Removal of aromatic compounds can be desirable for two different reasons. Either the aromatics have detrimental effects on the quality of the mixture they're in, or the aromatics are worth more if they're separated than if they're not. There are a number of examples, some of which have already been touched on.

a. Aromatic compounds in kerosene can cause unacceptable smoke points.

b. Kerosene range solvents that are aromatics-free or aromatics-laden have various industrial applications.

c. Separated benzene, xylene, and toluene have numerous chemical applications.

d. Removal of aromatics from heavy gas oil stocks can improve the lubricating oil characteristics.

Processes

The solvent recovery process is based on the ability of certain compounds to dissolve selectively certain classes of other compounds. In this case, certain solvents will dissolve aromatics but not paraffins, olefins, or naphthenes. The reasons the process works are complex and not worth examining here.

The other key feature of the solvents is that the solvent, with the aromatics dissolved in it, readily separates itself from the rest of the compounds. Take kerosene as an example and assume it has a lot of aromatic compounds in it. To half a beaker of kerosene add half a beaker of a solvent—in this case, liquid sulfur dioxide. After mixing, the liquid will separate into two phases with the kerosene on the bottom and the sulfur dioxide on top. The kerosene on the bottom will fill less than half the beaker. The sulfur dioxide, because the aromatic compounds have dissolved in it, will take up more than half the beaker.

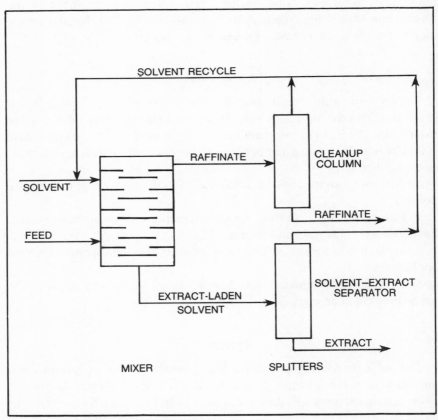

Fig. 17-1 —Solvent Recovery Process

If the sulfur dioxide is poured off, the aromatic compounds can be "sprung" by simple distillation. This two-step process is *batch solvent processing*.

Knowing how a simple batch process works, a continuous flow process is easy to conceptualize. In Fig. 17-1, a three column system is shown. The feed is introduced into the lower part of a vessel or column with a labyrinth of mixers inside. (Sometimes the mixers are mechanically moved, such as in a *rotating disc contactor*.) The solvent is introduced near the top. Almost all the solvent works its way towards the bottom of the vessel, dissolving the *extract* as it goes along. The rest of the hydrocarbon, which rises to the top, is called *raffinate*.

Two columns are used to handle the streams coming out of the mixer. One column just cleans up any solvent that may have followed along with the raffinate. The solvent is recycled to the mixer. The other column separates the solvent and the extract. The solvent from this column is also recycled to the mixer.

Some of the solvents used in various applications are listed below.

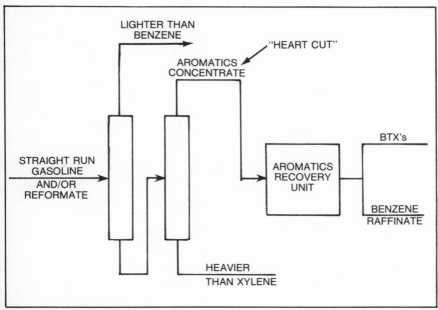

Fig. 17-2 —BTX Recovery

a. Kerosene treating: liquid SO₂, Furfural®

b. Lubricating oil treating: liquid SO₂ mixed with benzene, Furfural,® phenol, propane (separates paraffins from asphaltenes)

c. Gasoline: Sulfolane®, phenol, acetonitrile, liquid SO₂

Benzene Recovery

The most widespread application of solvent extraction is used in BTX recovery, especially for benzene. To make the process efficient, the feed to the process is pared down to an *aromatics concentrate* by making a *heart cut* on a reformate or straight run gasoline stream as shown below in Fig. 17-2. The aromatics concentrate then has a large benzene content, making the extraction process more efficient.

One bit of nomenclature is often misleading. *Benzene raffinate* contains no benzene. It is the leftover of the aromatics concentrate after the goodies (benzene) are removed. As a result, benzene raffinate is generally worth much less than aromatics concentrate, especially as a gasoline blending component.

XVIII.

ETHYLENE PLANTS

Soone hot, soone cold, nothing violent is permanent.
—"Petite Pallace," George Pettie

The closest chemical companies come to being petroleum refiners is in ethylene plants. That's the reason why so many large ethylene plants are built by integrated oil/chemical companies: they bridge the gap between the two. Ethylene plants are better called *olefins plants,* but are variously referred to as ethylene crackers (a misnomer), steam crackers (because steam is mixed with the feed) or crackers, with a suffix denoting the feed (ethane cracker, etc.).

Olefins plants can be designed to crack a number of feedstocks. They usually fall into the following categories:

> Ethane
> Ethane/Propane Mix
> Propane
> Butane
> Naphtha
> Gas Oil

The original olefins plants were built to produce ethylene, primarily to supply the growing appetite of the chemical industry for this basic building block. Of slightly less interest was propylene because there was a large pool of that hydrocarbon that could be stolen away from refinery alkylation plants if necessary. So, many of the early olefins plants were designed to crack ethane or ethane and propane because the yield of ethylene from ethane is very high. (See table on page 130.)

Later development of technology led to the use of the heavier feedstocks because of the larger availability of the feeds and because olefin plants cracking naphtha and gas oils produce a high octane gasoline blending component. A number of huge olefins plants as large as medium-sized refineries now are integrated into refineries and produce a significant amount of the gasoline.

129

OLEFIN PLANT YIELDS

Feed	Pounds Per Pound of Feed				
	Ethane	Propane	Butane	Naphtha	Gas Oil
Yield:					
Ethylene	0.77	0.40	0.36	0.23	0.18
Propylene	0.01	0.18	0.20	0.13	0.14
Butylene	0.01	0.02	0.05	0.15	0.06
Butadiene	0.01	0.01	0.03	0.04	0.04
Fuel Gas	0.20	0.38	0.31	0.26	0.18
Gasoline	—	0.01	0.05	0.18	0.18
Gas Oil	—	—	—	0.01	0.12
Fuel Oil	—	—	—	—	0.10

Refinery Interactions

The olefins plants provide a home for a number of junk streams in a refinery. For example, the dry gas stream from a cat cracker is usually sent to the fuel system, even though it contains ethane, ethylene, some propane, and propylene. But at the olefins plant, these components can be separated and put to higher value uses.

Some of the naphthas in the gasoline pool are low octane and of little value as a blending component. Benzene raffinate is an example. These streams can make an attractive feedstock for olefins plants, not only because of their low alternate value, but because they yield a higher octane component as a by-product of ethylene manufacture.

On the other side of the ethylene plant there are some complementary things, too. The butylenes and heavier feed can generally find a home in the refinery processes. By themselves they are not readily marketable, except to a refinery.

Process

The ethane/propane crackers are the simplest design, but they demonstrate the fundamentals. The ethane and propane can be fed separately or as a mixture to the cracking furnaces where the short residence time and high temperature, followed by a sudden cold quench, yield a high volume of ethylene (Fig. 18-1). But for the operating conditions and feeds, olefins plants are just plain old thermal crackers.

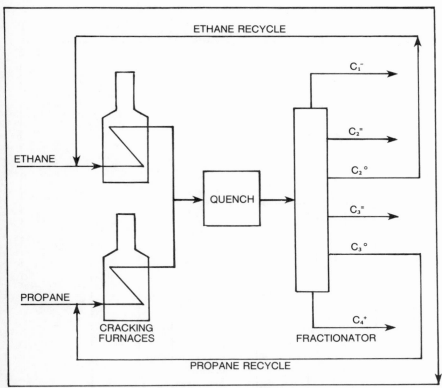

Fig. 18-1 —Olefins Plant: Ethane-Propane Cracker

In one pass through the cracking furnace, not all the ethane and propane disappear. So downstream, in the product fractionator, the ethane and propane are split out and recycled to the feed. Generally, the ethane is recycled to extinction, but some of the propane goes with the propylene. There are three commercial grades of propylene, depending on the amount of propane mixed in. Polymer grade propylene is 97-99% propylene; chemical grade is 92-95%; refinery grade is 50-65%.

The plants that crack the heavier liquids, naphtha, and gas oil create ethane on a once-through basis too. So, those olefins plants often have a furnace which is designed to handle the recycled ethane. Yields are usually shown as if the recycling took place, not the once-through basis.

A chemical stream that turns up here, but not in the refinery sector, is

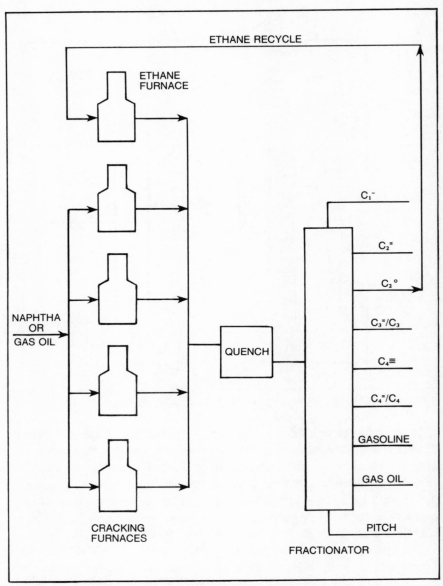

Fig. 18-2 —Olefins Plant: Heavy Liquids Cracker

butadiene (Fig. 18-3). This chemical is in the *diolefin* family and has two double bonds and the formula C_4H_6. The two double bonds make it particularly reactive and it is used in creating plastics and rubber compounds.

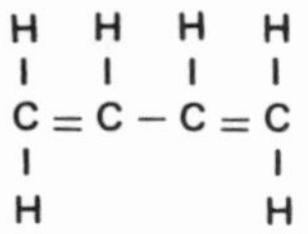

Butadiene

Fig. 18-3 —Butadiene

EXERCISES

1. How much feed, in barrels per day, is required to run a typical olefins plant producing a billion pounds per year of ethylene from the following feeds:

        Ethane (3.2 lb/gal)
        Propane (4.24 lb/gal)
        Naphtha (6.4 lb/gal)
        Gas Oil (7.3 lb/gal)

2. A company has an ethane/propane cracker with a capacity of 500 MM pounds per year of ethylene. Presently it is cracking a mixture of 70% (by volume) ethane and 30% propane, running the plant at capacity. Suddenly the propylene market turns sour, and the company wishes to produce only 20 MM lb per year of propylene. How much feed were they cracking, how much propylene were they producing, and how much ethane should they substitute for propane?

XIX.

SIMPLE AND COMPLEX REFINERIES

Aevo roussimo nostro, Simplicitas. (Simplicity, most rare in our age.)
—"Ars Amatoria," Publius Ovidius Naso

In the early 1980's, the businessmen and their economists who were running or analyzing refineries made an intellectual breakthrough. They recognized that the kinds of hardware inside different refineries were having profound effects on product prices, crude prices, and profitability. Since that time, close tracking of these variables has resulted in the terms *simple* and *complex* refineries, now routine nomenclature in refinery vocabulary.

Like so many words in the English language, the terms *simple* and *complex* are derived from only remotely related contexts. Twenty years before, in the 1960's, Wilbur L. Nelson developed a scale of *complexity factors*. He wanted to provide a general approach to determining the *investment cost* of various types of new refineries. Nelson's scheme related every possible major piece of refinery hardware to a crude distillation unit by means of capital cost. The crude unit was assigned a complexity of one. Each unit downstream of the crude unit had a factor in relation to its complexity and cost. For example, a cat cracker had a complexity of 4.0; it was four times more complex than crude distillation for the same amount of throughput.

To illustrate how the complexity factor is used, consider three types of refineries: fuel oil, gasoline, and petrochemical. The fuel oil refinery, shown in Fig. 19-1, is sometimes called a *hydroskimming refinery*. The name is derived from using hydrogen to upgrade the distillates skimmed from the crude. The hydrogen source is a catalytic reformer used to upgrade the naphtha to gasoline quality. The skimming operation leaves a large volume of heavy gas oil in the residue, so the yield of heavy fuel oil in a hydroskimming operation is quite large.

As for complexity, crude distillation is, by definition, 1. 0. The rest of the add-on complexity is calculated by multiplying the relative throughput

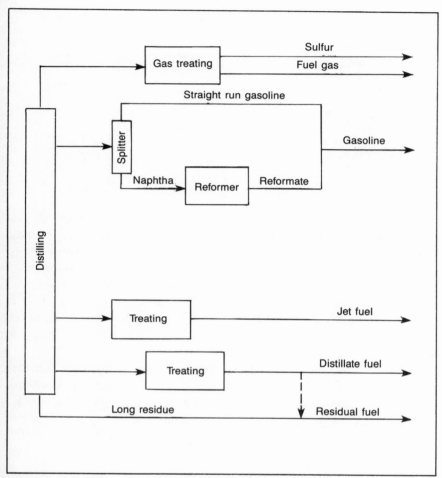

Fig 19-1 —Hydroskimming (or Simple) Refinery

of each unit by its complexity factor. For example, the cat reformer takes 15% of the crude unit output and has a complexity factor (calculated and supplied by Nelson) of 4.0; the add-on for the cat reformers is $0.15 \times 4.0 = 0.6$. Hydrotreating has a complexity of only 0.5 but treats 35% of the flow, so it adds 0.175. Adding up all the units' complexity factors, a hydroskimming refinery has a complexity of about 2.5, calculated as follows:

Table 19-1 Hydroskimming Refinery Complexity Calculation

	Complexity Factor	Throughput Ratio to Crude	Complexity
Crude distillation	1.0	1.0	1.000
Gas plant	0.5	0.5	0.250
Splitter	0.3	0.30	0.090
Naphtha hydrotreater	2.0	0.15	0.300
Catalytic reformer	4.0	0.15	0.600
Straight run gasoline treater	0.5	0.15	0.075
Kerosene hydrotreater	0.5	0.15	0.075
Distillate hydrotreater	0.5	0.20	0.100
			2.49

What about more complicated refineries that convert much of that residual fuel into gasoline and distillates? The complexity factor of these *gasoline refineries* goes up rapidly because the units added are very expensive. Adding the flasher (complexity factor of 2.0), cat cracker (6.0), hydrocracker (10.0), alkylation unit (11.0), and the auxiliary treaters and equipment puts the complexity of the gasoline refinery in Fig. 19-2 at 9–10.0. The residual fuel yield in one of these refineries is down in the range of 15–20%, while the gasoline yield is probably in the range of 45–55%.

At the extreme range of complexity are the refineries that have facilities to produce the high *value-added products* such as lube oils or petrochemicals. The complexity factors, i.e., the capital costs, are high. For aromatics recovery, the complexity factor is 33; for olefins, depending on the feed and downstream treating, 10–20. It isn't unusual to see a refinery with 10% petrochemicals yield (ethylene, propylene, butadiene, and aromatics) with a complexity of 16 or more.

The whole point of Nelson's analysis was to be able to draw (and use) the chart shown in Fig. 19-3. It relates refinery complexity to refinery cost and takes into account the economies of scale for refinery size. The scale on the left is just an index. It must be calibrated to the current inflation adjusted cost of construction for a crude distilling unit. The rest of the chart shows the relationship of crude distillation throughput (the normal indicator is for "refinery size") to complexity (now the more sophisticated engineering approach to size).

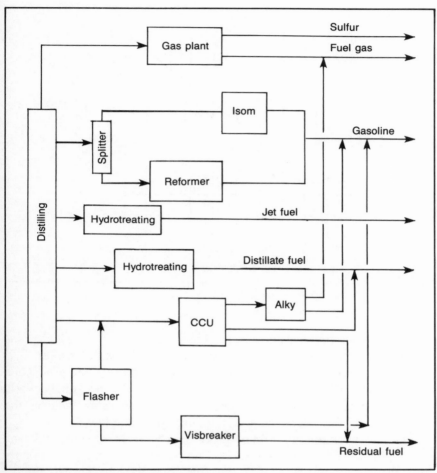

Fig 19-2 —Gasoline (or Complex) Refinery

The New Look at Complexity

Nelson's construct was created in an era of refinery expansion. The interest was in the cost of the next refinery to be built. But in the 1980's, the interest (concern) was not in expansion but in operations; not in capital

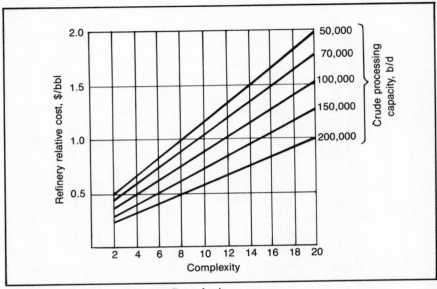

Fig 19-3 —Refinery Cost vs. Complexity

costs but in operating margins. You might say the idea of complexity was reformed to accommodate that interest; the catalyst was the market.

The key concept is uncomplicated: the more higher-valued products a refinery makes, the more it can pay for raw material. But in a volatile world where raw material costs, product prices, and profit margins fluctuate, the implications of that statement are profound. An example will help.

Take a hydroskimming refinery processing West Texas sour crude and selling products into the spot market. Compare that to a gasoline refinery doing the same thing. Table 19-2 illustrates some hypothetical economics. It is clear that if West Texas sour crude is available to either refinery at a price of $28/bbl, and if the product prices are those shown, then the gasoline refinery will make more money than the hydroskimming refinery. The more complex refinery has a bigger operating margin than the less complex (simple) one. That's easy enough.

But change one variable and a slightly different message emerges in Table 19-3. Instead of West Texas sour crude at $28/bbl, substitute a very heavy crude like Mayan (from Mexico) at a price of $26/bbl. The distillation curve and other characteristics of Mayan crude are different than West Texas sour, but the hardware is still the same in each refinery. So

Table 19-2 Refining $28/bbl West Texas Sour Crude	Hydroskimming Refinery			Gasoline Refinery		
	% Volume	Unit Cost or Price	Revenue or Total Cost, $	% Volume	Unit Cost or Price	Revenue or Total Cost, $
West Texas sour crude	100	$28	28.00	100	$28	28.00
Gasoline	30	32	19.60	50	32	16.00
Jet fuel	10	32	3.20	19	32	6.08
Distillate fuel	20	31	6.20	17	31	5.27
Residual fuel	35	27	9.45	20	27	5.40
Refinery fuel (gain)	5	—	—	(6)	—	—
Total outturn	100		28.45	100		32.75
Operating costs	100	1	1.00	100	3	3.00
Operating margin	100		(0.65)	100		1.75

the yields are quite different, and look what happens to the operating margins! The simple refinery makes almost as much money as the complex refinery.

There are a few important observations at this point:

1. Mayan crude has about the same profitability in either type refinery, simple or complex.*

Table 19-3 Refining $26/bbl Mayan Crude	Hydroskimming Refinery			Gasoline Refinery		
	% Volume	Unit Cost or Price	Revenue or Total Cost, $	% Volume	Unit Cost or Price	Revenue or Total Cost, $
Mayan crude	100	$26	26.00	100	$26	26.00
Gasoline	10	32	3.20	25	32	8.00
Jet fuel	5	32	1.60	5	32	1.60
Distillate fuel	20	31	6.20	25	31	7.75
Residual fuel	60	27	16.20	50	27	12.15
Refinery fuel (gain)	—	—	—	(5)	—	—
Total outturn	100		27.20			29.50
Operating costs	100	1	1.00	100	3	3.00
Operating margin	100		0.20	100		0.50

*The nomenclature in the text is different than the tables (simple vis-a-vis hydroskimming and complex vis-a-vis gasoline) just to get you to make the mental association.

2. Simple refiners can't afford to run West Texas sour. The acquisition cost is too high for the product slate that is produced. Complex refiners make enough light products to give a positive margin.

3. It might be that the complex refiner has bid up the price of West Texas sour crude so that the simple refiner can't afford it.

4. The yields from Mayan are such that the complex refiner has hardly any room to bid it up and take it away from the simple refiner. The price of Mayan is already so high at $26/bbl that neither simple nor complex makes much margin ($0.20–$0.50/bbl).

While that third observation might seem profound, it might also be a gross exaggeration. The situation may instead be that there is so much Mayan crude around that it fills up the complex refineries and some has to be run in the simple refineries, too. (Hence, the same margin in both types of refinery.) Furthermore, the product demands (and prices) are such that additional crude (West Texas sour) has to be run in complex refineries to meet the light oil demands.

Deciding which of those conditions prevails often isn't possible by looking only at crude oil margins for any specific point in time. There are too many variables to pin it down. But as you'll see, changes over time can tell a story.

First, though, one more example. Suppose the relationship between the light oil (gasoline, jet fuel, and distillate fuel) prices and the resid prices, i.e., the *resid/light oil differential,* changes. In Tables 19-2 and 19-3, the average resid/light oil differential is about $4.67. Say for some reason it increases to $8.67/bbl; light oils increase in price by $3/bbl, resid falls off by $1/bbl. Table 19-4 shows the results.

At first glance it might seem that both types of refiners are better off—all the operating margins are positive and larger. But closer examination indicates that the simple refiner is more vulnerable. The spread between

Table 19-4 Refinery Operating Margins		
	Hydroskimming	*Gasoline*
West Texas Sour		
4.67 spread	$(0.35)	$1.75
8.67 spread	0.70	4.13
Mayan		
4.67 spread	0.20	0.50
8.67 spread	0.65	1.65

simple and complex operating margins has increased, too. That gives the complex refiner more room to maneuver, perhaps bidding up the crude prices to the point of eliminating any margin for the simple refiner.

Will the price be bid up? If the simple refiner can no longer afford to refine either crude, will his shutdown affect the product supply/demand balances? Will that affect the resid/light oil differential? And will that ultimately provide an incentive for the simple refiner to start-up again?

Industry's Model

All those questions are only a sample of what is asked every day as crude oil prices and product prices and differentials change. To cope, some rigor has been introduced to the analysis, and it shows up even in the industry press. In general, refineries are classified as simple or complex, although analysts often use "very complex" as well. The simple and complex refineries are about the same as Nelson would define them.

Simple Refinery—crude distillation, hydrotreating of middle distillates, cat reforming of naphtha.

Complex Refinery—simple refinery plus a cat cracker, alkylation plant plus gas processing.

Very Complex Refinery—complex refinery plus either an olefins unit or a residual reduction unit such as a coker.

Any crude run through these three refineries will have higher light oil yields in the very complex refinery than the simple. Take West Texas sour, for instance:

Table 19-5 Complexity and Yields			
	% Yield		
	Simple	Complex	Very Complex
Gasoline	30	50	65
Jet fuel	10	19	20
Distillate fuel	20	17	25
Residual fuel	35	20	0
Fuel (gain)	5	(6)	(10)

Another crude will have its own set of yields from each refinery. In the typical industry model, the yield for each crude for each type refinery has to be individually calculated. In that way, the incentives to refine what and where can be observed. Typically, that's done by looking at the results of the calculations two different ways: a snapshot of how the whole industry looks now and a look at how parts of it operate over time.

The Snapshot

The computer and communications technology are available now to permit a weekly (or even daily) calculation and publication of a comparison of buyers' incentives to acquire individual crude oils. Two locations have to be specified: the point where the product sales are priced and the point where the crude oil is costed. The products are usually priced wherever the crude oil is refined, e.g., the U.S. Gulf Coast, Rotterdam, Singapore. The crude can be costed at that same point or it can be costed at the producing point. In the latter case, either the products or the crude have to be debited for the crude oil transportation cost. Table 19-4 has U.S. Gulf Coast sales netted back to the production fields for simple and complex refineries for prices prevailing on the day of the calculation in some destination market.

	Table 19-6 Complexity vs Netbacks					
	Product Revenue Yield			*Netback at Source*		Crude
Crude Oil	*Simple*	*Complex*	*Freight*	*Simple*	*Complex*	*Oil Price*
Arab Light	30.50	33.50	1.20	29.30	32.50	29.00
Arab Heavy	28.50	30.50	1.30	27.20	29.20	26.00
Mexican Mayan	26.00	29.00	1.40	24.60	27.60	25.00
Nigerian Light	35.10	37.00	1.60	33.50	35.40	30.00
North Sea Brent	34.60	36.60	1.50	33.10	35.10	30.00

From such a list, which is usually much longer, you can determine whether the typical simple or complex refiner has any interests in buying up these crudes. By comparing the *netbacks* to the crude costs, you can tell which ones are the bargains and which ones are the dogs.

The Industry Track

To capture the dynamics of the markets, a plot like that in Fig. 19-4 can be used. The yields of simple and complex refineries for a given crude (two locations specified again) are plotted over time, together with the "offi-

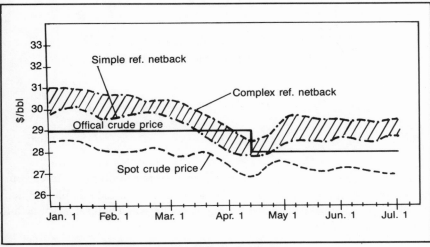

Fig 19-4 —Crude Oil Prices and Refinery Netbacks

cial" price and the spot price. In that plot, the pressures building over time can be tracked. Pressures to change either the product prices or the crude prices can be observed. In Fig. 19-4, it looks like the decline in product prices (the netbacks) drove down first the spot crude prices, then the official crude prices.

Even when you are all done preparing the lists or the plots, the answers are all not obvious. But plotting or comparing simple and complex netbacks vs crude costs gives you a good start.

CRUDE OIL, CONDENSATE, AND NATURAL GAS LIQUIDS

Get your facts first, and then you can distort them as much as you please.

—Mark Twain

The raw materials coming into a refinery from the oil patch have some names that are not too descriptive. Sometimes they sound all alike. What's the difference between crude and condensate; between natural gasoline and natural gas liquids?

Oil Patch Operations

A simplistic look at what goes on in a producing oil field is shown in Fig. 20-1. The underground accumulations of hydrocarbon can be in several

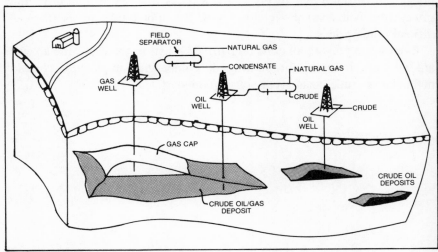

Fig. 20-1 —Oil Patch Operations

forms. The well on the left has tapped a reserve that has a *gas cap* at the top. Production from that well is *gas well gas*. This hydrocarbon mixture is predominantly methane, but 15-20% of the material could be heavier hydrocarbons, ranging up to the gas oils.

As the gas comes out of the well and is transported, it cools down and more of the heavier hydrocarbons liquefy. The mixture is introduced into a vessel called a *field separator*. This vessel is sometimes referred to as a "wide spot in the line." What happens is that in the larger space the pressure drops, and whatever light ends have dissolved in the heavier liquids flash. *Natural gas* is drawn off the top of the separator. The bottoms are *condensate*.

The well in the middle has tapped the same formation, but the production is crude oil with some dissolved natural gas. At the surface, another field separator will separate this mixture into natural gas and crude oil.

The well on the right has tapped another formation in which there is only crude oil. No separator facilities are necessary.

Further Processing. Crude oil and condensate can be collected and transported by truck, tank car, or pipeline. Natural gas almost always moves by pipeline. If there's no pipeline around, the natural gas might have to be reinjected back into the ground to keep the associated crude oil production coming.

When crude oil is handled by pipeline, often a number of different grades of crude oil from different formations are brought together in a single mix. If it is convenient, condensate is mixed in with the crude as well.

Despite the fact that the natural gas has gone through field separators, there may still be some hydrocarbons heavier than the methane-ethane range. The natural gas may be processed in a gas processing plant for the recovery of the *natural gas liquids* (NGL's). The NGL's consist of ethane, propane, butanes, and *natural gasoline*, which is a mixture of pentanes through C_9's or C_{10}'s. Sometimes the natural gasoline and butane content may be large enough that during the cold winter months they may condense in the natural gas transmission lines. Pumps may be made inoperative and operations interrupted. In this case the natural gas *must* be processed in a gas plant to make it transportable and marketable by removing the NGL's.

Gas Processing Plants

Four basic schemes may be used to recover the NGL's.

1. Refrigeration Plant
2. Cryogenic Refrigeration
3. Lean Oil Absorption
4. Dry Bed Adsorption

In the *refrigeration plant,* the liquids-laden gas is cooled to a temperature somewhere in the range of 15°F down as low as minus 40°F. At the lower temperature about 70% of the ethane, 90% of the propane, and all of the butanes and heavier liquefy and can be separated for fractionation.

The so-called *deep-cut* ethane recovery operation is done in *cryogenic* refrigeration plants. In this operation, an apparatus called a turbo-expander is used to drop the temperature of the liquids-laden gas to −150°F to −225°F. Under these conditions, 90–95% of the ethane and all the propane and heavier drops out of the gas.

The older gas processing plants are the *lean oil absorption plants,* of the same design as the refinery gas plants. A plain lean oil absorption plant will recover about 70% of the propane and all the butanes and heavier. If a refrigerated lean oil is used, all of the propane can be recovered plus 50-75% of the ethane.

Dry bed adsorption is an interesting process that is used when a gas is to be processed only for dew point control (removing only the heavier liquids that might condense in transit). Many gas sales contracts require that the dew point (the temperature at which droplets form) be no higher than 15°F at the pipeline pressure of 800 psi. Depending on the gas stream, that may require all the natural gasoline and some of the butanes to be removed.

Certain porous materials like activated charcoal, silica gel, and alumina have the power to cause large amounts of vapors to condense on their surface. Since the liquid ends up on the surface, the process is called *ad*sorbing, in contrast to *ab*sorbing where the liquids end up inside. After a sufficient amount of liquids (in this case NGL's) is condensed out of the gas, the process is shut down—or directed to another vessel of adsorbent. The liquids are then driven off the adsorbent with steam, collected, and condensed.

Dry bed adsorption will recover 10–15% of the butanes and 50–90% of the natural gasoline.

Transport and Disposition

The market for most of the oil patch production is refineries. As mentioned above, most of it moves in pipelines to the refinery centers. It has often been convenient and efficient to use the crude oil as a carrier for not only the condensate, but the natural gasoline and butane as well. Crude oil with butane and/or natural gasoline injected in it is called *volatiles-laden crude*. Sometimes the vapor pressure limits on the pipeline determine the maximum amount of butane that can be injected in the crude.

Propane and ethane are often handled differently than butane and natural gasoline. Since propane is a marketable product, it is often separated at the plant and moved away by truck, tank car, or pipeline. Ethane, on the other hand, is pretty much limited to pipeline movement because of its vapor pressure, density, etc. Sometimes it is moved segregated, sometimes mixed with the other NGL's. In that case the mixture of natural gasoline and lighter is separated from the gas in the gas plant, but is not further fractionated. It is moved as a mixture called *raw make*. Eventually, the raw make will probably end up being fractionated at a plant site closer to the market or a more convenient redistribution point.

XXI.

FUEL VALUES—HEATING VALUES

If you can't stand the heat, get out of the kitchen.

—Harry S. Truman

Many of the economic forces on the various petroleum products are tied to the amount of heat the individual products yield when they are burned. Indeed, the choice of which streams to use as refinery fuel must take into account the market value of the streams and the heating value.

Thermal Content

When a hydrocarbon is burned, two things happen: a chemical reaction takes place and heat is generated. Typically, the chemical reaction is a transformation of the hydrocarbon and oxygen into water and carbon dioxide:

$$CH_4 + 2O_2 \rightarrow CO_2 + 2H_2O + heat$$
$$2C_6H_6 + 15O_2 \rightarrow 12CO_2 + 6H_2O + heat$$

The amount of heat given off by the reaction is unique to each type of hydrocarbon. The normal measure of heat in the petroleum industry is the BTU, which stands for British thermal unit.

Definition. The amount of heat required to raise the temperature of one pound of water 1°F is equal to one BTU.

A table of the heating values for some commercial petroleum products is given below:

Product	Higher Heating Value
Natural gas	1000–1050 BTU/scf
Ethane	66,000 BTU/gal
Propane	91,600 BTU/gal
Butane	1,3,300 BTU/gal
Distillate fuel	140,000 BTU/gal
No 6. fuel (2.5% sulfur)	153,000 BTU/gal
No 6. fuel (0.3% sulfur)	151,500 BTU/gal

There are two types of heating values. The so-called higher heating value represents the gross amount of heat given off by the chemical reaction (the heat of combustion). The lower heating value takes into account what happens to the water created. A portion of the heat of combustion is absorbed to vaporize the water. Not all that heat is recovered or usable as the heat goes through a furnace and out a stack. As a general rule, the heavier the fuel, the lower the ratio of hydrogen to carbon so the less water formed during combustion. Typical results of this inability to recapture the heat of vaporization are measured by the thermal efficiency, the recoverable heat divided by the gross heating value.

Product	Thermal Efficiency, %
Natural gas	84
Propane	85
Distillate fuel	88
Coal	90

Competitive Fuel Value Nomogram

A handy way to relate the values of the various products and their heating values is shown in the nomogram in Fig. 21-1. That peephole diagram first appeared in the *Oil & Gas Journal* in 1972 and was updated to put higher scales on it in 1977 and then again in 1980.

As an example of nomogram use, assume you wish to find the fuel values equivalent to $2/MM BTU natural gas. First, go to the vertical scale on the right and find the horizontal line that intersects it at $2. Next, read along the horizontal line, right to left, the intersections on the other fuels scales: 13.2 cpg for ethane, 18.3 cpg for propane, 20.8 cpg for butane, etc.

Similarly, the value of any fuel can be equated to any other fuel by extending a straight line from the vertex on the left through the first fuel's value until it intersects the second fuel's scale.

For example, by laying a straightedge from the vertex through the high sulfur No. 6 fuel oil scale at $10/bbl, you can see the equivalent values are $34/ton for lignite, 22 cpg for distillate fuel, 14 cpg for propane, and $1.57/MM BTU for natural gas.

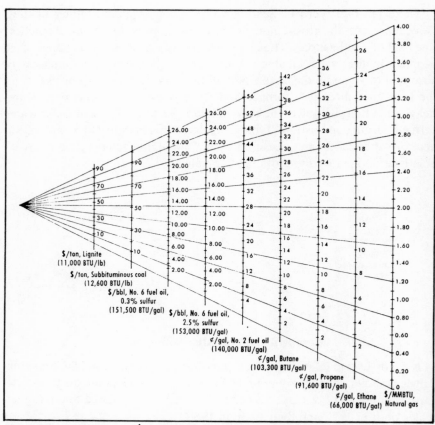

Fig. 21-1 —Equivalent Fuel Values for selected hydrocarbons

ANSWERS AND SOLUTIONS

What is the answer? . . . In that case, what is the question?
—Alice B. Toklas, What is Remembered, Gertrude Stein

CHAPTER II

1. (a) Calculate the cumulative percent volumes and plot.
   (b) The Oklahoma Sweet has 9.7%. The Heavy California crude only starts to boil somewhere between 260° and 315°, and so it does not even have the full range of naphtha. The volume is only 4.2%.
2. Because of the peculiar formula for API gravity, you can't add them and divide by two. You have to convert the API gravities to specific gravities, add them, divide by two, then convert them back to API gravity:

$$11° = 0.9930 \text{ sp. gr.}$$
$$50° = \underline{0.7796} \text{ sp. gr.}$$
$$1.7726 \div 2 = 0.8826 = 28.15°$$

CHAPTER III

1. a. cooler
   b. batch, continuous
   c. bubble cap
   d. bubble cap . . . downcomer
   e. end point . . . initial boiling point
   f. decreases
   g. decreases . . . increases
2. There are several ways to solve the problem. One simplifying assumption you *can* make is that the distillation curve is a set of straight lines between the cut point/percent volume plots.

Then calculate a composite distillation curve by figuring the volume in each cut from the two crudes.

	Volume-M B/D
IBP-113	31.5
113-260	25.8
260-315	35.1
315-500	21.5
500-750	37.6
750-1000	20.0
1000+	28.5
	200.0

The next step can be done either by algebra or by plotting the distillation curve. The algebraic method is as follows. The IBP of the SRLGO is 525. That means that some of the 500-750° cut goes into jet fuel. How much? Let X be the amount. Then,

$$\frac{525-500}{750-500} = \frac{X}{37.6}$$

$$X = 3.76$$

That means the 525–750 cut has in it:

$$37.6 - 3.76 = 33.84 \text{ M B/D}$$

To get only 20 MB/D, starting at 525°, let Y = the end point. Then,

$$\frac{Y - 525}{750 - 525} = \frac{20.0}{33.84}$$

$$Y = 658°$$

CHAPTER IV

1. a. distillation . . . cracking
   b. lower
   c. lower
   d. light flashed distillate, heavy flashed distillate, and flasher bottoms
   e. increase, lower, lower

2. a. To get 35 M B/D flasher bottoms, start at the "bottom of the barrel" on the distillation curve and work up. In problem 2, Chapter III, the volume of 1000° + material is 28.5 M B/D. So, all that must be flasher bottoms. The 750°-1000° cut (20 M B/D) will cover the remainder of the requirement of 35.0-28.5 = 6.5 M B/D. Calculate as follows:

> Let X be the IBP of the flasher bottoms. (Also known as the flasher temperature and the end point of the flasher tops.) Then,

$$\frac{1000 - X}{1000 - 750} = \frac{6.5}{20}$$

$$X = 918.75°$$

b. The flasher tops come from the 750−1000° cut, and go from 800−918.75.

> Let Y = Volume of flasher tops.
> Then,

$$\frac{Y}{20} = \frac{918.75 - 800.0}{1000 - 750}$$

$$Y = 9.5 \text{ M B/D}$$

CHAPTER V

1. Any way you build C_3H_8, the three carbons are connected to each other in the same fashion: two of them have three hydrogens and a carbon attached, one has two hydrogens and two carbons. The fact that you can draw on paper one of the carbons attached at right angles to the others doesn't mean that in nature the molecule looks any different.

2. There's only one structure for isobutane, a branch off the middle of three carbons.

Isobutane

There are two isomers of pentane. Isopentane has a branch off either of the two inside carbons in a string of four. It doesn't matter which one; they're symmetrical if you turn them around. The other isomer, neopentane, has four carbon branches off the inside carbon.

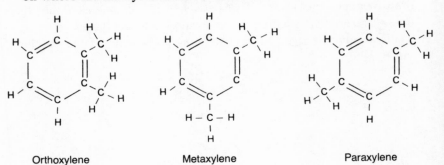

Isopentane   Neopentane

The one isomer of butylene can best be depicted by drawing the carbons slightly cock-eyed. Then, no matter how you rotate it, it looks the same.

Isobutylene

3. Paraffins, Olefins, Naphthenes, and Aromatics.
4. The three types of xylene, called *para, meta,* and *orthoxylene,* depend on where the methyl radicals are attached.

Orthoxylene   Metaxylene   Paraxylene

5. It doesn't matter where a single methyl radical is attached since the benzene ring is symmetrical.

CHAPTER VI

1. a. catalyst . . . hydrocarbon.
   b. coke or carbon . . . air . . . CO . . . CO_2.
   c. heavy straight-run cuts . . . gasoline.
   d. distilling column . . . flasher.
   e. olefins.
   f. cycle oil . . . recycled to extinction.
2. From the question on Flashing, flasher tops, with cut points $800-918.75°$ F, are 9.5 M B/D. The cut points of the straight-run heavy gas oil must be the end point of the straight-run light gas oil, 658° F, (from the question on Distilling) and the initial boiling point of the straight-run residue, 800° F. The nasty job of calculating the volume of this $658°-800°$ F cut goes as follows:

Volume of $500-750°$ F = 37.6	(Distilling problem)
$500-525°$ F = 3.76	(Distilling problem)
$525-658°$ F = 20.0	(Distilling problem)

   Therefore,
   Volume of $658°-750°$ F = 37.6−20−3.76
   $$= 13.84$$

Volume of $750-1000°$ F = 20.0	(Distilling problem)
of $800-1000°$ F = 16.0	(Flasher problem)

   Therefore,
   Volume of $750-800°$ F = 4.0

   Therefore,
   Volume of straight-run heavy gas oil =
   $$658°-800° \text{ F} = 13.84 + 4.0 = 17.84$$

Yield of cat light gas oil is 12%; cat cracker feed is the straight-run heavy gas oil and flasher tops, so the cat light gas oil volume is 0.12 × (17.84 + 9.5) = 3.28 MB/D.

3. (1) Decrease the initial boiling point or (2) increase the end point of the cat light gas oil; (3) increase the feed to the cat cracker by increasing the crude distilling column feed rate or (3) changing the cut points of the straight-run heavy gas oil and the flasher tops; (4) increasing the cycle oil volume by lowering the end point of the cat heavy gas oil; (5) change the operating conditions in the cat cracker reactor, or (6) the regenerator to change the cracking yields.

CHAPTER VII

1. a. sats . . . cracked
   b. lean oil . . . fat oil
   c. sponge oil
2. Methane—Refinery fuel
   Ethane—Refinery fuel or chemical feed
   Propane—Commercial fuel or chemical feed
   Normal butane—Motor gasoline blending
   Isobutane—Alkylation feed
   Propylene—Alkylation
   Butylenes—Alkylation
   Ethylene—Refinery fuel or chemical feed
   Hydrogen—Hydrotreating
3. One of the forms of butylenes and the isobutylene boils at lower temperatures than normal butane; the other two normal butylenes boil at temperatures higher. So the butylenes get split from each other to get the normal butane out.

   One distilling column configuration is in Diagram 1. Several other schemes are possible as well. The important factor is understanding what goes overhead, what doesn't.
4. Diag. 2

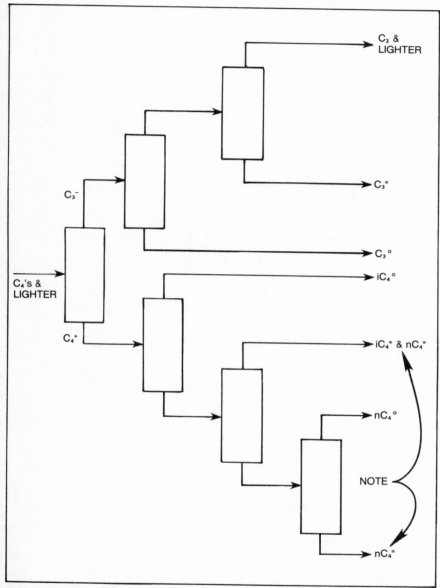

Diagram 1—Distillation Column set-up to segregate cracked gases

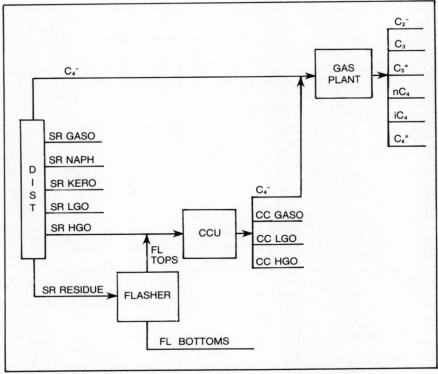

Diagram 2—Crude Distilling column, flasher, cat cracker and gas plant

CHAPTER VIII

1. a. cracking
   b. sulfuric acid or hydrofluoric acid
   c. cooler, reactor, acid separator, caustic wash, and fractionators
   d. isoheptane and isooctane
   e. octane number . . . vapor pressure . . . gasoline blending component
   f. shrinkage

2. $iC_4°$ for $C_3=$: $3000 \times .25 \times 1.6 = 1200$ B/D
     for $C_4=$: $3000 \times .30 \times 1.2 = \underline{1080}$ B/D
                                        2280

less the 300 B/D in the feed is   1980 B/D

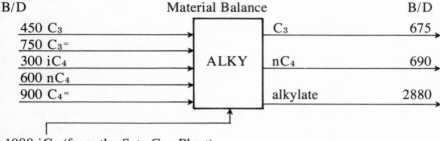

B/D   Material Balance   B/D

450 C_3	C_3	675
750 $C_3=$		
300 iC_4	**ALKY**	nC_4	690
600 nC_4		
900 $C_4=$	alkylate	2880

1980 iC_4 (from the Sats Gas Plant)

Yields
   $C_3°$: $(3000 \times 0.15) + (3000 \times 0.25 \times 0.3)$ $=$ 675 B/D
   $C_4°$: $(3000 \times 0.20) + (3000 \times 0.30 \times 0.1)$ $=$ 690 B/D
alkylate: $(3000 \times 0.25 \times 1.8) + (3000 \times 0.30 \times 1.7)$ $=$ 2880 B/D

                                          Feed $=$ 4980 B/D
                                  Outturn $=$ 4245 B/D

CHAPTER IX

1. a. octane
   b. platinum
   c. aromatics
   d. hydrogen
   e. methane, ethane, propane, and butanes
   f. hydrogen
   g. yield . . . octane number . . . butanes and lighter

2. The values of the yield are computed as follows.

From the chart:	91 Oct.	95 Oct.	100 Oct.
Reformate Yield % Volume	88.0	08.7	75.0
C4 and lighter yield, % Vol.	8.7	10.4	25.2
Reformate Value, ¢/gal.	100.0	104.0	109.0
C4 and lighter value, ¢/gal.	50.0	50.0	50.0
Reformate value	8800.0	8800.8	8175.0
C4 and heavier value	435.	525.0	1260.0
Total Value	9235.0	9333.8	9435.0

The total value keeps going up, only slightly, but up, as the octane number is raised despite the falling yield of reformate. So it makes sense to increase severity.

3. Parafins to Isoparaffins:

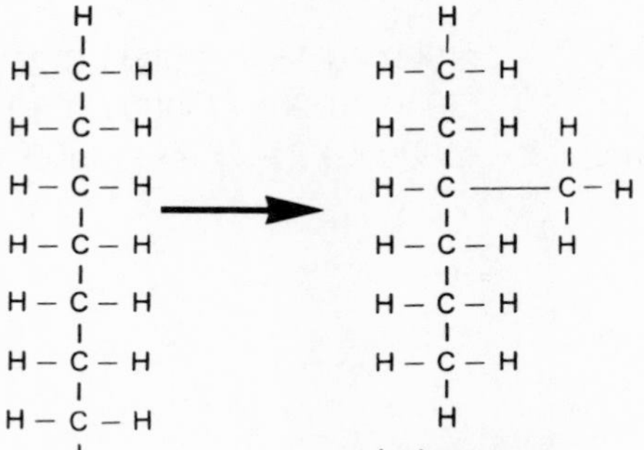

Normal heptane

Isoheptane

Paraffins to Naphthenes:

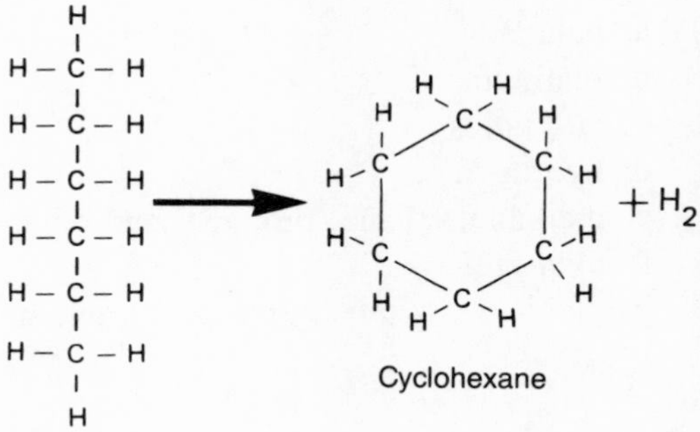

Normal hexane

Cyclohexane

$+H_2$

Naphthenes to Aromatics

Cyclohexane   Benzene $+ \quad 3H_2$

Naphthenes Crack to Butanes and Lighter

Cyclopentane $+ \; 2H_2$ Butane   Methane

Side Chains Crack off Aromatics

Xylene $+ \; 2H_2$ Benzene   Methane $+ \; 2CH_4$

4. Platinum plus reformate gives platformate

5. Diag. 3

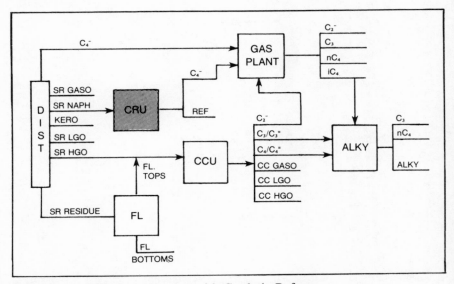

Diagram 3—Refinery Operation with Catalytic Reformer

CHAPTER X

1. a. Catalyst
   b. Coke and Residue
   c. Cooled with a recycle stream
   d. "Cooked" until it cokes
   e. Low or poor
   f. Sponge . . . needle
   g. Cracked . . . olefins
   h. Furnace, distillation column, and reactor . . . drum

2. Flasher bottoms are 35 MB/D

$$\text{Coke Yield} = \frac{35,000 \times 350 \times 0.30}{2000} = 1837.5 \text{ Tons/Day}$$

3. Diag. 4

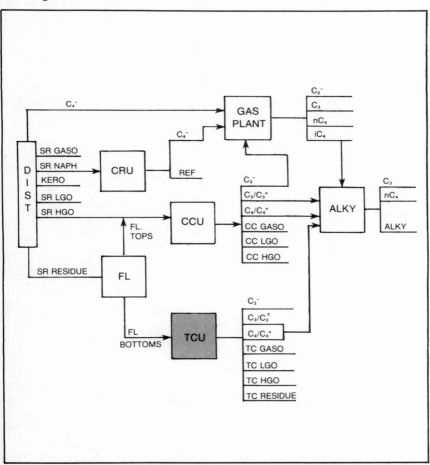

Diagram 4—Thermal Cracker added to Refinery Process

CHAPTER XI

1.

	Hydro-cracking	Cat Cracking	Thermal Cracking
Feed	gas oils	heavy straight-run gas oil	CCHGO flasher bottoms
Process promoter	catalyst, hydrogen	catalyst	heat
Product quality	paraffinic, naphthenic	naphthenic, aromatic olefinic	paraffinic, naphthenic, aromatic olefinic

2. Cat cracker gas oils make good feed to hydrocrackers. Hydrocrackate makes good feed to reformers.

3. Diagram 5

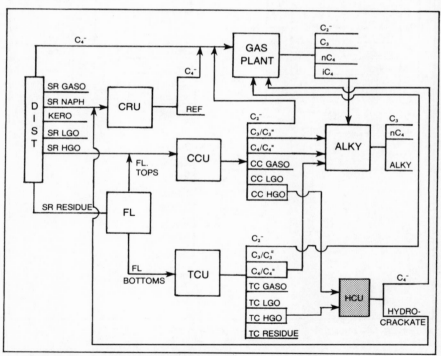

Diagram 5—Refinery Flow Diagram with Hydrocracker

CHAPTER XII

1. *Vapor pressure* is a measure of the surface pressure necessary to keep a liquid from vaporizing.

RVP stands for Reid Vapor Pressure and is the numerical result of measuring vapor pressure using Reid's method.

Power stroke is the downward motion of a piston that occurs after ignition.

Vapor lock is the phenomenon of insufficient gasoline flow from a fuel pump due to its inability to pump liquid-gas mixtures.

Pressuring agent is the hydrocarbon, usually normal butane, used to bring gasoline blends up to an acceptable vapor pressure.

Knocking is the preignition of gasoline in a cylinder during the compression stroke.

RON and *MON* are measures of octane numbers under conditions simulating mild and severe conditions, respectively.

Leaded gasoline is gasoline that has TEL or TML added to boost the octane number.

Octane enhancement is the reaction to the addition of lead as measured by MON or RON.

2.

	Barrels	RVP	RON–0	RON–3.17
SR Gasoline	2,730	11.1	66.4	83.5
Reformate	2,490	2.8	94.0	100.8
Hvy Hydrocrackate	6,100	1.7	79.0	92.2
CC Gasoline	3,600	4.4	92.3	95.8
Subtotal	14,920	4.25		
Normal butane	X	52.0	93.0	102.5

$$(14,920)(4.25) + 52.0X = (14,920 + X)(12.5)$$
$$X = 3,116$$

	Barrels	RVP	RON–0	RON–3.17
Blend, including the butane	18,036	12.5	84.2	94.6

Plotting 84.2 and 94.6 on the octane number versus lead chart (Diagram 6) indicates that 4.8 g/gal need to be added, which exceeds the EPA limit of 3.17 g/gal.

If the heavy hydrocrackate is reformed, a low octane gasoline blending stock is replaced by a high octane component. Assuming the yield of 85%, the reformate volume goes up to 7,675, the hydrocrackate yield goes to zero.

	Barrels	RVP	RON-0	RON-3.17
SR Gasoline	2,730	11.1	66.4	83.5
Reformate	7,675	2.8	94.0	100.8
CC Gasoline	3,600	4.4	92.3	95.8
Subtotal	14,005	4.83		
Normal butane	X	52.0	93.0	102.5

$$(14,005)(4.83) + 52.0 \ X = (14,005 + Z)(12.5)$$
$$X = 2,719$$

	Barrels	RVP	RON-0	RON-3.17
Blend, including the butane	16,724	12.5	89.0	97.2

The lead requirement in Diagram 6 for the new blend is only 2.2 g/gal, below the legal limit.

CHAPTER XV

	Hydrotreating	DEA Removal	Claus Plant	SMR
H_2S	P	F, P	F, I	—
S	—	—	P	—
CO	—	—	—	I
CO_2	—	—	—	I
CH_4	—	—	—	F, P
H_2	F	—	—	P
O_2	—	—	F	—
SO_2	—	—	I, P	—

P–Product
F–Feed
I–Internal Stream

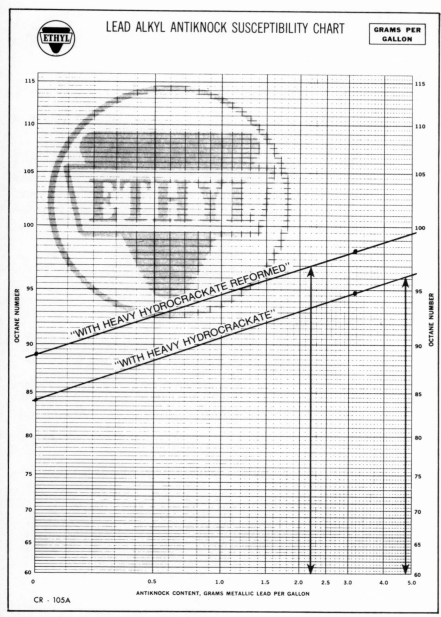

Diagram 6—Octane Enhancement Chart

CHAPTER XIX

1. Ethane: $1 \times 10^9 \div (0.77 \times 3.2 \times 42 \times 365) = 26$ MB/D
   Propane: 50 MB/D
   Naphtha: 44 MB/D
   Gas Oil: 50 MB/D

2. Let X equal the feed in b/d of the 70/30 mix. To make 500 MM lb of ethylene, the feed rate is as follows:

   Calculate the weighted average ethylene yield from the ethane and propane. The yield from ethane is X times the percent ethane times the pounds per gallon ethane times the yield of ethylene from ethane times 42 gal per barrel times 365 days per year. Same for ethylene from propane.

From ethane	From propane	Ethylene
$(X)(0.7)(3.2)(.77)(42)(365)$ +	$(X)(0.30)(4.24)(0.40)(42)(365)$	= 500×10^6

$$X = 14.6 \text{ mb/d}$$

The propylene production was:

From ethane	From propane	Propylene
$(14.6)(0.7)(3.2)(42)(365)(0.1)$ +	$(14.6)(0.3)(4.24)(42)(365)(0.18)$ =	56.5×10^6

To find out how much ethane and propane must be cracked to make 500 MM lb of ethylene but only 20 MM lb of propylene, use simultaneous equations:

Let Y equal the ethane feed rate in b/d
    Z equal the propane feed rate in b/d

Then:

	From ethane	From propane
ethylene:	$500 \times 10^6 = (Y)(3.2)(42)(365)(0.77)$ +	$(Z)(4.24)(42)(365)(0.40)$
propylene:	$20 \times 10^6 = (Y)(3.2)(42)(365)(0.01)$ +	$(Z)(4.24)(42)(365)(0.18)$

$$500 \times 10^6 = 37{,}773 \text{ Y} + 26{,}000 \text{ Z}$$
$$20 \times 10^6 = 490 \text{ Y} + 11{,}700 \text{ Z}$$
$$Y = 12.4 \text{ MB/D ethane}$$
$$Z = \underline{1.2 \text{ MB/D propane}}$$
$$13.6 \text{ MB/D of 91/9 mix}$$

INDEX

There are men who pretend to understand a book by scouting through the index: as if a traveler should go about to describe a palace when he had seen nothing but the privvy.

"On The Mechanical Operation of the Spirit," Jonathan Swift

(Italicized page numbers are the primary references)